Coastal and Estuarine Studies

Series Editors:
Malcolm J. Bowman   Christopher N.K. Mooers

Coastal and Estuarine Studies

52

Bo Barker Jørgensen and
Katherine Richardson (Eds.)

Eutrophication in Coastal Marine Ecosystems

American Geophysical Union

Washington, DC

Eutrophication in Coastal Marine Ecosystems / Bo Barker Jørgensen
    and Katherine Richardson (eds.).
            p.   cm. -- (Coastal and estuarine studies ; 52)
        Includes bibliographical references and index.
        ISBN 0-87590-266-9
        1. Marine eutrophication. 2. Coastal ecology. I. Jørgensen, Bo Barker.
    II. Richardson, Katherine. III. American Geophysical
    Union. IV Series.
    QH91.8.E87E87 1996   96-29236
    574.5'2638--dc20   CIP

ISSN 0733-9569

ISBN 0-87590-266-9

Copyright 1996 by the American Geophysical Union, 2000 Florida Avenue, NW, Washington, DC 20009, U.S.A.

Printed in the United States of America.

CONTENTS

PREFACE

Over the last 10–20 years, eutrophication has become generally acknowledged as an environmental threat for many coastal marine areas. Nevertheless, most of our knowledge of the effects of eutrophication on aquatic ecosystems is derived from limnological studies. There are fundamental differences between marine, brackish, and freshwater systems that lead to different responses to eutrophication. Therefore, the aim of this book is to provide the reader with a synthesis of our current understanding of eutrophication in coastal marine areas and the biological/ecological responses to it.

This book evolved out of the interaction between scientists from a number of different disciplines involved in the Marine Research Program sponsored by the Danish Ministry of Environment and Energy during the period 1988–1994. This research program was established in the wake of considerable public attention given to oxygen depletion events which repeatedly occurred during late summer in Kattegat bottom waters during the 1980s. In 1987, the Danish Parliament adopted measures to reduce the nitrogen and phosphorus input from Denmark to the surrounding marine waters by 50% and 80%, respectively. Acknowledging the need for more research into the ecological effects of eutrophication, the Parliament elected at the same time to fund the Marine Research Program.

While conceived within and partially supported by a Danish research program, this book is designed as a general text dealing with coastal eutrophication in temperate waters. A unique feature of the book is the integration of a general description of the chemical and biological processes in coastal marine ecosystems with a discussion of the effects of eutrophication, drawing examples mainly from Danish waters. The reader will find a generic presentation of the processes (hydrographic and atmospheric) that are responsible for the delivery of nutrients to coastal marine waters, as well as for the effects of eutrophication on pelagic and benthic processes. Two papers deal specifically with eutrophication of the very shallow waters found at the land-sea interface. The final papers treat the problem of collating and presenting relevant scientific results in a form which is useful for the administrator who is dealing with eutrophication issues. Several of the papers are followed by 'case studies' in which specific projects carried out within the Marine Research Program are described. We intend these case studies to be illustrative of some of the general principles presented in the main papers and to serve as models for other workers planning similar studies in other areas. So that others can profit most from the experiences gained in connection with the Marine Research Program, we have attempted to be as candid as possible with respect to the successes and failures which resulted from the research projects described in the case studies.

We thank all of our friends and colleagues who have helped to produce this book—especially our colleagues around the world who served as anonymous reviewers for the chapters presented here. Finally, we acknowledge that the dream which led to this book could not have been realised without the financial support of the Danish Environmental Protection Agency or the administrative support of many of the Agency's personnel—in particular Jørn Kirkegaard and Jesper Andersen, who supplied wise guidance and support throughout the production of the book.

Katherine Richardson
Bo Barker Jørgensen

1

Eutrophication:
Definition, History and Effects

Katherine Richardson and Bo Barker Jørgensen

1.1. Introduction

"Eutrophication" of coastal marine waters is a much discussed topic – both in the scientific and popular literature. Unfortunately, however, few of those who use the term take the time to reflect upon the process that is actually being discussed and a number of different uses (definitions) of the term have come into common practice. In fact, the term eutrophication has its roots in two Greek words – *eu* meaning "well" and *trope* which means "nourishment". Eutrophication can be defined as the process of changing the nutritional status of a given water body by increasing the nutrient resources.

The most frequent use of the term and that which is employed throughout this book is to describe increases in the input of mineral nutrients (primarily nitrogen and phosphorus) to a particular water body. However, as pointed out by Nixon [1995], it is not only nitrogen and phosphorus that determine the nutritional level (= trophic state) of a water body. He suggests that eutrophication be defined as "an increase in the rate of supply of organic carbon to an ecosystem". This proposed definition has the advantage that it allows us, at least to some degree, to quantify the trophic state of the water bodies with which we are working on the basis of a single property.

Organic carbon input to a system comes either via primary production within the system (autochthonous carbon) or via external input to the system (allochthonous carbon). There are few good measurements of allochthonous carbon input to marine systems. However, the primary production in many marine systems is well measured. Thus, Nixon [1995] has proposed a classification scheme which describes four trophic states (oligotrophic, mesotrophic, eutrophic and hypertrophic) for

Eutrophication in Coastal Marine Ecosystems
Coastal and Estuarine Studies, Volume 52, Pages 1–19
Copyright 1996 by the American Geophysical Union

TABLE 1.1.

	Organic carbon supply (primary production) $g\ C\ m^{-2}\ y^{-1}$
Nixon's (1995) classification scheme for marine waters.	
Oligotrophic	≤ 100
Mesotrophic	100–300
Eutrophic	301–500
Hypertrophic	>500
Rodhe's (1969) classification scheme for lakes	
Oligotrohic	7–25
Eutrophic (natural)	75–250
Eutrophic (polluted)	350–700

marine coastal and estuarine ecosystems based on the primary production occurring there. A similar scheme was developed for fresh waters by Rodhe [1969]. These two classification schemes are presented in Table 1.1.

Most marine systems can be classified as being from oligotrophic to eutrophic in their natural states. The Sargasso Sea, where the water column is thermally stratified throughout the year and mineral nutrients become concentrated in subsurface waters where light does not penetrate, is a good example of an oligotrophic marine system. This book, however, is concerned with coastal regions of temperate shelf seas and these are typically classified as being meso- or eutrophic. In such seas, seasonal changes in solar heating and wind activity acting together periodically destroy the stratification of the water column and ensure transport of nutrient-rich deep water into the upper layer of the water column where photosynthesis can take place. Meso- and eutrophic seas have a higher fish production than oligotrophic seas – a fact that has led to the quite serious suggestion that fisheries in some waters may be improved by the addition of nutrients (see section 1.6).

Eutrophication can occur as a result of natural processes. It occurs, for example, via "upwelling" – in regions where the local hydrographic conditions lead to transport of nutrient-rich deep water (from the light limited aphotic zone) to nutrient- poor but light-rich surface waters of the "euphotic" (or "photic") region of the water column. For the most part, however, when we speak of eutrophication, it is "cultural" eutrophication or that which is associated with anthropogenic activities which is of interest. It should be noted that both in popular and scientific literature it has now become common to (incorrectly) extend the application of the term eutrophication to include not only the process of increasing the nutritional status of a region but also to encompass the *effects* of this enrichment.

Whether or not eutrophication is occurring naturally or as a result of anthropogenic activities, it affects the ecosystem in which it is occurring in the same manner. The most immediate effect of eutrophication in most systems is an increase in the plant (phytoplankton and/or macrophyte) biomass. The plant biomass present at any given point in time is a function of plant production (growth) and loss (via grazing, sedimentation, mortality and dilution) processes. The production process, growth, is, in turn, a function of light and nutrient availability and (to a lesser degree) temperature. Thus, any of these three factors alone or in combination can be limiting for the production of plant biomass in the marine environment at any point in time. When a nutrient is limiting or colimiting, a change in nutrient availability through eutrophication can affect the plant biomass in an ecosystem.

1.2. Nutrient Limitation

Most of our understanding of nutrient limitation of phytoplankton comes from laboratory studies carried out on unialgal cultures (see, for example, review by Hecky and Kilham [1988]). As a result, there is now a reasonable understanding of the mechanisms of nutrient limitation at the cellular level. However, when considering entire ecosystems and the potential effects of eutrophication upon them, it is necessary to consider the effects of nutrient limitation at levels higher than that of the individual cell.

For some purposes, for example, it can be relevant to consider nutrient limitation at the level of populations [e.g., Goldman et al., 1979; and discussion by Howarth, 1988]. For other purposes, it may be relevant to consider nutrient limitation of community net primary production or net production of the entire ecosystem [see Howarth, 1988 for examples]. The effects of nutrient enrichment at these levels of the ecosystem are much less understood than those occurring at the cellular level and much of the research being carried out on eutrophication and its effects today focuses on the effect of nutrient enrichment or reduction at the community or ecosystem level. It is interesting to note that much of the apparent disagreement found in the literature concerning nutrient limitation and the causes/effects of eutrophication can be explained by the fact that comparisons are, in some cases, being attempted between studies carried out at different hierarchical levels within the ecosystem.

The eutrophication effects we recognize are those resulting from a net increase in plant biomass within the system. Such an increase in biomass can only occur through an increase in "new" production as described by Dugdale and Goering [1967]. In the cases they described, the external input of nitrogen in the form of nitrate is predicted to yield an increase in phytoplankton biomass. These authors also identified a form of production which they called "regenerated" based upon the recycling of nitrogen within the euphotic zone (see chapters 4 and 5, this volume).

Regenerated production does not lead to a net increase in biomass within the system. Standard primary production measurements are not able to distinguish between new and regenerated production. Thus, an increase in the measured primary production rate does not necessarily imply an increase in biomass but can simply be the result of an increase in nutrient recycling rates within the euphotic zone.

The stimulation of plant biomass which can result from eutrophication has potential ramifications for energy flow and can give rise to a shift in the balance of naturally occurring processes within the ecosystem. It is these changes in the balance between the different processes occurring within the system that give rise to the effects that are popularly associated with eutrophication (hypoxia, algal blooms, changes in species diversity and abundance, etc. – see sections 1.6–1.9).

1.3. Cultural Eutrophication: History

It is believed that man has been aware of the effects of eutrophication since before the beginning of written history. The fact that the Aztecs of Tenochtitlan chose not to dispose of their excretory products in the nearby lake but, instead, spread them on land suggests that they were aware of and concerned about cultural eutrophication [Vollenweider, 1992]. In addition, many believe that it was an algal bloom caused by cultural eutrophication that is being referred to in the Bible:

> *... and all the waters that were in the river were turned to blood.*
> *And the fish that was in the river died; and the river stank, and the Egyptians could not drink of the water of the river...* (Exodus 7: 20–21).

The cultural eutrophication of European lakes (leading to algal blooms and oxygen deficiency) has certainly been acknowledged as a problem since the early 1800s [see Vollenweider, 1992]. As a result, the study of eutrophication's impact on ecosystems has developed almost exclusively within the domain of limnology. Understanding of eutrophication and its effects is now well developed within fresh-water systems and considerable success has been achieved here with eutrophication control based on models designed to identify "acceptable" levels of eutrophication (see chapter 11).

However, it is only relatively recently that cultural eutrophication of marine waters has been perceived as a potential environmental threat [e.g., Rosenberg, 1985; Gray, 1992; Nixon, 1995]. The lack of earlier concern over eutrophication of marine areas is largely explained by the fact that marine areas are characterized by a greater water exchange with surrounding water bodies than landlocked lakes. As a result of this water exchange, there is a greater capacity for nutrients entering a marine system to be diluted and transported away from the input site than there is in most lakes.

1.4. Eutrophication in Marine versus Fresh-water Systems

In spite of this generally greater ability for dilution of nutrients in marine systems, the fundamental ecosystem response to eutrophication is similar in fresh and marine waters. Thus, much can be learned about marine eutrophication by studying the limnological literature. However, there are essential differences between these two types of systems that prevent us from simply applying knowledge gained through limnological studies to the marine environment.

The models developed to identify "acceptable" or "tolerable" levels of eutrophication in fresh-water systems, for example, are for the most part not immediately applicable to marine areas. Many of these models are based on Vollenweider's [1976] empirical (regression) model for lake eutrophication which includes the flushing rate (annual rate of water exchange) and sedimentation rates for both nitrogen and phosphorus. The more variable hydrodynamic properties and water exchange rates in most marine as opposed to lake systems and the fact that rates of and processes leading to N and P sedimentation are generally not well documented for marine systems makes the immediate application of Vollenweider-type models difficult [Gray, 1992; see also chapter 11].

Another difference between fresh and marine systems is that phytoplankton are generally believed to be phosphorus limited in the former and nitrogen limited in the latter. There is a lively debate within the scientific community as to whether this is, in fact, the case and, if so, why this should be [see, for example, Smetacek et al., 1991; Codispoti, 1989; Hecky and Kilham, 1988; Howarth, 1988; Smith and Hollibaugh, 1989] and it is not the intention to actively enter this debate or to review the problem here.

Hecky and Kilham [1988] have, however, pointed out that the relative elemental requirements of marine and fresh water phytoplankton appear to be similar. On the basis of a comparison of the relative elemental composition of phytoplankton and that of average ocean and river waters (Table 1.2), they argue there are more potential elemental candidates for limiting phytoplankton growth in marine than in fresh water. In fresh water, they identify only the relative presence of phosphorus as being lower than or similar to phytoplankton elemental composition. In marine waters, however, nitrogen, silicon and phosphorus are all identified as being present in relative amounts similar to or less than those found in phytoplankton. Thus, their analysis suggests that P limitation of fresh-water phytoplankton is the most likely while the situation with respect to marine phytoplankton is more uncertain.

TABLE 1.2. Relative elemental composition of algae (normalized to total dissolved P (molar basis)) compared to relative mean composition in river and ocean waters. From Hecky and Kilham [1988].

Element	River	Algal	Ocean
C	738	102	1,000
N	28(21)	11.1	13
Si	146	96	43
K	26	1.3	4,434
P	1.0	1.0	1.0
S	146	0.54	12,000

It should be noted that the Hecky and Kilham analysis also included a comparison of the relative elemental composition of phytoplankton and marine and fresh waters with respect to metals. For fresh waters, only iron and cobalt were identified as being present in relative quantities similar to those found in phytoplankton. Iron, zinc, copper, manganese and cobalt were all found to be present in relative quantities in marine waters that were similar to or less than those reported for phytoplankton.

While this comparison of the relative elemental composition of phytoplankton relative to marine and fresh water does not establish which element(s) is (are) limiting for phytoplankton growth, it does suggest that the apparent differences reported with respect to phytoplankton nutrient limitation in marine and fresh waters may have their basis in the fundamental differences in chemistry between the two types of aquatic environments.

The geochemical cycling of N and P is also affected by the basic chemical differences between marine and fresh water. While these processes are still not well understood, it is clear, for example, that marine sediments have a lower P-binding capacity than fresh-water sediments. This is at least partly due to differences in iron chemistry in the two sediment types that can be related to the greater availability of sulfate/sulfide in marine than in fresh water.

Thus, although both marine and fresh-water systems can respond to eutrophication by increasing plant biomass, the differences in marine and fresh-water chemistry alone suggest that the processes surrounding eutrophication in marine systems may be quite different from those occurring in fresh water. As a result, it is argued here that the study of eutrophication in marine environments is a unique scientific discipline which is related to but distinct from the study of eutrophication in fresh-water systems.

1.5. Cultural Eutrophication in Marine Waters: Extent of the Problem

Examples of marine areas where cultural eutrophication is now perceived to be a threat can be found along the edges of all continents with the exception of Antarctica. Nixon [1990] identifies the regions of the coastlines of Europe, North and South America, Africa, India, south-east Asia, Australia, China and Japan as all suffering from undesirable effects of cultural eutrophication.

It is areas of reduced water exchange, for example,

- semienclosed seas such as the Baltic and Adriatic,
- fjords with sills that restrict water exchange,
- lagoons, bays and harbors,

which typically exhibit evidence of adverse effects of cultural eutrophication. In addition, however, some coastal areas, particularly those under river influence where mixing with open sea waters is limited are also areas of concern. A good example of such an area is the south-east "corner" of the North Sea (The German Bight) where the Elbe, Rhine and Weser Rivers enter the North Sea. Here, signs of hypoxia and anoxia along the sea bottom have been observed.

Most of the public and political interest in eutrophication is triggered by the actual manifestation of visible effects of cultural eutrophication in coastal waters and, from the discussion above, it is clear that such effects are found along nearly all marine coastlines. While this concern with visible changes in the local environment is understandable, it may well turn out that the real importance of marine eutrophication on a global scale is its potential influence on geochemical cycling.

Many authors [e.g., Smayda, 1990; Richardson and Ærtebjerg, 1993; Billen et al., 1991; Conley et al. 1993] have pointed out that cultural eutrophication alters the relative availability of nutrient elements in coastal waters. In particular, it has been emphasized that while cultural eutrophication increases the delivery of N and P to coastal waters, it has very little influence on Si delivery. Not all algal groups have a requirement for Si. Therefore, it has been suggested that changing the relative availability of Si in relation to N and P will discriminate against phytoplankton with a silicon requirement (diatoms) [e.g., Billen et al., 1991 and references therein].

Indeed, in their 1993 review, Conley et al., conclude that there is evidence for changes in the relative availability of Si and/or changes in the relative occurrence of different functional algal groups which are consistent with those predicted under conditions of reduced Si availability from marine coastal regions around the world. It should be noted here that, while most eutrophication-related discussion has focused on the total phytoplankton community or biomass, the relative occurrence of different functional groups in the phytoplankton community is not purely of academic interest. Changes in the phytoplankton community's species composition can have profound effects on the structure of the rest of the food web (see section 1.7 and chapters 4 and 5, this volume). In addition, conditions favoring flagellate rather than diatom development may increase the probability of occurrence of toxic/noxious phytoplankton blooms [e.g., Smayda, 1990].

The stimulation of phytoplankton growth/biomass by eutrophication also has a potentially important role to play in the carbon cycling taking place in the world's oceans. Considerable scientific effort is presently being directed toward elucidation of the carbon cycle in the sea. A better understanding of these carbon cycling processes is clearly important in terms of the discussion concerning global warming (greenhouse effect) predicted on the basis of the anthropogenically induced global increase in production of CO_2. However, the quantitative role of the oceans in global carbon cycling is still not clear [see for example, Mantoura et al., 1991].

Nevertheless, on the basis of the oceans' size alone, it is intuitively obvious that carbon cycling here must be a dominating factor in terms of global carbon cycling. The quantitative global impact of a eutrophication induced increase in algal biomass (productivity) on carbon fixation, sedimentation and burial has not yet been established. However, it is clear that, by its very nature, marine eutrophication can, at least potentially, affect carbon cycling processes occurring in the sea. Thus, while it is the visible effects of eutrophication occurring locally that attract attention to marine eutrophication, it is possible that the real importance of the phenomenon may be found at the global level.

1.6. Eutrophication: Enrichment Phase

As indicated earlier, most systems will respond to eutrophication by generating a greater biomass of plant material. Sedimentation of this organic material will increase food availability for benthic organisms and result in changes in benthos biomass. The increase in organic material in the pelagic and benthic systems increases the food availability for fish and can result in an increase in fish biomass. Gray [1992] calls this the "enrichment phase" in his general scheme of eutrophication effects (Figure 1.1) and it is the existence of this initial enrichment phase that gives rise to proposals aimed at increasing fish biomass through eutrophication. Indeed, some authors [e.g., Tatara, 1991; Larsson et al., 1985] have argued that the increase in fisheries yield which has been recorded in coastal seas in recent centuries may be a direct result of anthropogenic eutrophication. Changes recorded in the growth rate of herring in the Baltic have also been attributed to eutrophication [Hansson and Rudstam, 1990].

When considering proposals to increase fisheries yield through eutrophication, however, it is important to remember that eutrophication will only lead to an increase in fish biomass if all levels of the food chain leading up to fish are limited by

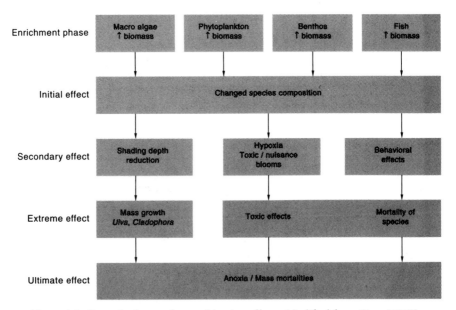

Figure 1.1. General scheme of eutrophication effects. Modified from Gray [1992].

nutrient or food availability. This concept is, perhaps, best illustrated by using a terrestrial analogy in which the fish are represented by a field of cattle. The grass which they eat would then represent plankton. If growth of the cattle is limited by food availability then fertilizing the grass (eutrophication) will relieve their starvation and lead to an increase in biomass. However, there will come a level of fertilization where some other factor than food availability limits the biomass production of the cattle (water, physiological reproduction rate, etc.). Above this level of fertilization, no benefit in the yield of cattle will be achieved.

Just as there is a level above which increased fertilization will not increase the biomass of cattle in a field of grass, there is a level of nutrient availability over which no "benefit" will result in the yield of fisheries. Thus, it is only ecosystems in which the biomass of fish (or the animals they eat if they do not feed directly on phytoplankton) is limited by food availability that can be enriched to produce a greater fish biomass via eutrophication.

1.7. Eutrophication: Initial and Secondary Effects

In addition to increasing the biomass at one or more trophic levels, eutrophication can lead to a change in species composition. Eutrophication induced changes in species composition of benthic organisms are, perhaps, the best well-described (see chapter 8). However, there has been increasing awareness in recent years concerning the importance of the size structure of the phytoplankton community in controlling energy transfer in the food web (see, for example, Cushing [1989]; Kiørboe [1993]; and chapter 4 in this volume).

Any eutrophication induced change in the species composition of the phytoplankton community which leads to a change in size structure of the phytoplankton community will potentially affect energy flow in the entire ecosystem. Thus, eutrophication can, at least in theory, play an important role in dictating whether the higher trophic levels in a given system are dominated by marketable fish or by jellyfish [e.g., Fisher, 1976; Greve and Parsons, 1977]. Little is actually known about the effects of eutrophication on the size structure of the phytoplankton community under various conditions but this is an area that warrants further research.

The increased plant biomass resulting from eutrophication will give rise to secondary effects on the ecosystem. Such effects include a reduction in the depth to which light can penetrate in the water column and a subsequent reduction in the depth distribution of benthic plants (see chapter 9).

1.8. Eutrophication and Algal Blooms

Another secondary effect of eutrophication can be exceptional "blooms" (rapid growth and high biomass accumulation of a single phytoplankton species) of toxic or nuisance phytoplankton. Blooms of the prymnesiophyte, *Phaeocystis sp.* in the

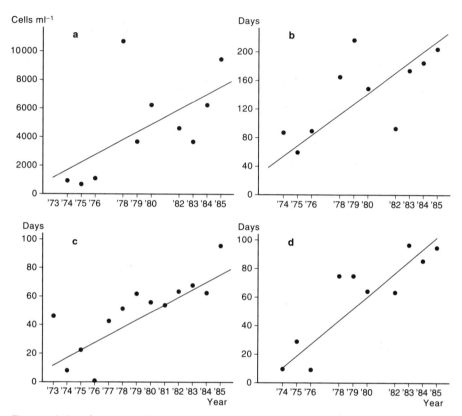

Figure 1.2. Development in the annual occurrences of *Phaeocystis* in the southern North Sea during the period 1973–1985. a) Annual averages of numbers of *Phaeocystis* cells ml⁻¹, based on equally weighted monthly averages to account for irregular sampling over the year ($r = 0.61$, $p < 0.05$); b) Numbers of days per year with more than 100 *Phaeocystis* cells ml⁻¹ ($r = 0.73$, $p < 0.01$); c) Duration of the *Phaeocystis* spring peak (period with more than 1000 cells ml⁻¹) ($r = 0.89$, $p < 0.01$); d) Numbers of days per year with more than 1000 *Phaeocystis* cells ml⁻¹ ($r = 0.77$, $p < 0.01$). From Cadée and Hegeman [1986.]

southern North Sea have, for example, often been attributed to cultural eutrophication and there seems little doubt that bloom occurrences of this organism have increased in recent years concomitantly with increasing cultural eutrophication of this area (Figure 1.2).

It should be noted, however, that the relationship between cultural eutrophication and blooms of toxic/nuisance phytoplankton is far from completely understood and cultural eutrophication is not always implicated in bloom formation. On the basis of the variation observed in the occurrence of dinoflagellate cysts in sedimentological sections taken from a region on Bornholm, Denmark (Figure1.3), it has been argued that phytoplankton blooms were occurring 140 million years ago. The workers studying these sediments [Noe-Nygaard and Surlyk, 1988; Noe-Nygaard

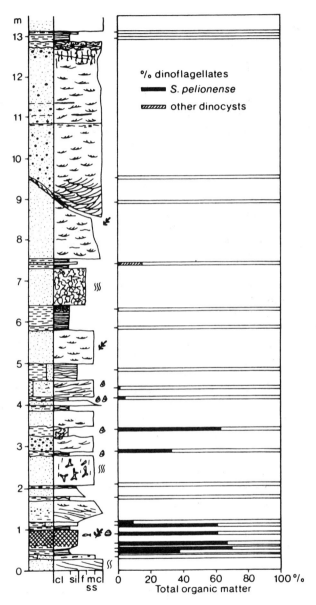

Figure 1.3. Sedimentological section made from a region on Bornholm, Denmark, showing the monospecific occurrences of the dinoflagellate *Sentusidium pelionense* in beds with census populations of the bivalve, *Neomiodon angulata*. The relative proportion of dinoflagellate cysts in relation to the total organic matter is indicated. It is not possible from such a section to establish cause and effect between the dinoflagellate blooms and mass mortalities of *Neomiodon*. However, the repeated coincidence of these events suggests that the bivalve mortalities may have been caused by the dinoflagellate blooms. From Noe-Nygaard et al. [1987.]

et al., 1987; Piasecki, 1984, 1986] have also argued that there is evidence in this fossil record for mass mortalities of bivalves occurring contemporaneously with the large occurrences of dinoflagellates.

In addition, records of the early European explorers of North America describe very clearly the effects of eating shellfish from an area where toxic phytoplankton were present [e.g., Hallegraeff, 1993]. Shellfish collection has been banned for much of recent history off the coast of Alaska because of the threat of toxic algal blooms. Thus, toxic phytoplankton blooms have been demonstrated to occur in regions not suffering from cultural eutrophication.

Scientists [e.g., Maestrini and Granéli, 1991] and media alike were quick to blame a 1988 bloom of the prymnesiophyte, *Chrysochromulina polylepis*, in the Skagerrak/Kattegat on cultural eutrophication. However, evidence is accumulating that suggests that the occurrence of the bloom cannot be directly related to an increase in the total nutrient availability in 1988 relative to other recent years. Gray [1992] refers to evidence that the biomass of *Chrysochromulina* recorded could have been supported by "normal" stocks of nitrogen in this area.

In addition, Heilmann et al. [1994] have examined water column characteristics relating to phytoplankton distribution and productivity during the bloom with those found during the same time period in the years preceding and following 1988. They demonstrated that there were significant differences in the amount of chlorophyll recorded in surface waters as well as in the relative proportion of total water column chlorophyll found in a subsurface chlorophyll peak as opposed to at the surface during the different years. However, the situation observed during the *Chrysochromulina* bloom did not represent an extreme with respect to the amount of chlorophyll present or its relative position in the water column. These authors were unable to demonstrate a significant difference in the total water column primary production occurring in the time period examined in the different study years (1987–1993).

Thus, what was interesting about this bloom was the fact that a single toxic species came to dominate the phytoplankton community. Cultural eutrophication may be implicated as a causative agent here through, for example, a change in the relative availability of nitrogen and phosphorus [Granéli et al., 1993; Edvardsen et al., 1990]. However, the mechanism(s) of this potential interaction between the *Chrysochromulina* bloom and cultural eutrophication has (have) not yet been identified.

It has been argued [i.e., Smayda, 1990] that cultural eutrophication has increased the frequency of toxic and nuisance blooms and this is probably the case, at least in some areas. Certainly, in the Seto Inland Sea (Japan) there has been a demonstrable decrease in the frequency of "red tides" (water discoloration due to excessive phytoplankton growth) after control measures were taken to reduce the chemical oxygen demand (COD) of the waters discharging into this sea [Prakash, 1987]. However, it is not yet clear whether or not cultural eutrophication in some manner "selects" for toxic organisms (through a change in the relative abundance of toxic species) or whether the apparent increase in frequency of nuisance and toxic phytoplankton blooms is a result of the fact that these organisms represent a subset of all phytoplankton and that there has been an increase in total phyto-plankton biomass. Thus, considerable work still remains to be done in order to quantify the relationship between nuisance and toxic phytoplankton blooms and cultural eutrophication.

1.9. Eutrophication and Hypoxia

Perhaps the most serious and certainly most widely discussed local effect of excess eutrophication is the "hypoxia" (oxygen concentration lower than at air saturation ("normoxia")) that it can cause in the water column. Hypoxia is another secondary effect of the increase in phytoplankton biomass that results from eutrophication. The increased plant production increases the delivery of organic material to the bottom via sedimentation and the microbial processes associated with the decay of this organic material consume oxygen causing a drawdown in the oxygen concen-tration in surrounding waters. In extreme cases, this hypoxia is so severe that "anoxia" (zero oxygen concentration) occurs in the bottom waters, hydrogen sul-fide production in the sediment is strongly enhanced and mats of sulfur-reducing bacteria may form upon the sea bottom (see chapters 6 and 7).

Different units of oxygen concentration in sea water are used in the chapters of this book. The concentration units of $\mu mol\ l^{-1}$, $ml\ l^{-1}$, and $mg\ l^{-1}$ have the following conversion factors: 0.0320 mg μmol^{-1}, 0.0224 ml μmol^{-1}, and 1.428 mg ml^{-1}. Where oxygen concentration is given as % saturation, this is the % of air satu-ration at the ambient temperature and salinity at atmospheric pressure. In chapter 8, the physiologically more relevant term, oxygen tension (i.e., oxygen partial pres-sure), is used with the unit of torr (mm mercury), which is $1/760$ of a standard atmosphere. At air saturation (21% oxygen), the oxygen tension in sea water is 160 torr. The conversion factor between oxygen tension and concentration varies with salinity and temperature.

As the most important sources of oxygen to marine waters operate in the surface waters (the air/water interface and phytoplankton photosynthesis), it follows that the regions in which hypoxia is most likely to develop are those where surface and bottom waters are separated from one another by a steep density gradient or "pyc-nocline". The density of water is determined from its temperature and salinity char-acteristics (see chapter 3). In some cases, it is temperature and, in others, salinity that dominates in determining the water column's density profile. Thus, the pres-ence of a pycnocline can be identified by the presence of either a "thermocline" (steep temperature gradient) or a "halocline" (steep salinity gradient) in the water column.

The most important sources of oxygen to bottom waters are down-mixing (usually wind-driven) from the surface through the pycnocline and the advection of bottom water containing oxygen from surrounding seas. Thus, natural variations in water exchange processes and weather conditions (which control wind-mixing of surface and bottom water) play an important role in determining the development and intensity of hypoxic events in a given region.

When oxygen consumption exceeds oxygen delivery to the bottom water, hypoxia will result. Again here, however, it is important to emphasize that hypoxia can result from natural eutrophication processes as well as cultural eutrophication. Gray [1992] cites an example of hydrogen sulfide (H_2S) production (which only takes place under anoxic conditions) occurring naturally under the thermocline in upwelling areas off Peru. What is interesting about cultural eutrophication in this respect is that it is causing an increase in the intensity and the distribution of hypoxia in coastal waters.

Biological systems are, of course, dependent on oxygen. Thus, even small changes in oxygen availability can give rise to physiological changes in the organisms found in the system (see chapter 8). The first response to hypoxia that can be observed at the population level is a behavioral one. Many fish, for example, will respond by fleeing an area of reduced oxygen concentration. This often leads to increased catches especially for fishermen using gears placed at a fixed geographic position (the fact that greater numbers of fish are moving greater distances increases the probability of them meeting a net). Thus, slightly reduced oxygen availability can give a boom in the local fishery.

Hypoxia can also increase catches of less mobile organisms. The Norway lobster (*Nephrops norvegicus*), for example, lives in burrows in the sea bed and is fished in the Kattegat and Skagerrak by bottom trawling. This trawling, however, only catches individuals that are out or on their way out of their burrows. Those individuals safely tucked away in their underground homes will not be caught. The oxygen availability in these burrows, however, is considerably lower that in the free-moving waters above the sea bottom. During periods of low oxygen availability in the bottom waters, the oxygen conditions in the burrows can become intolerable and the lobsters are forced up to the waters covering the sea bottom in order to survive (see chapter 8). This makes them more vulnerable to trawling and catches of Norway lobster show a clear increase in response to hypoxia (Figure 1.4).

As oxygen becomes less available in bottom waters, the less tolerant species begin to die – thus changing the structure of the bottom faunal community (and increasing the organic material available for microbial degradation). In severe cases, bottom waters may become anoxic causing die-off of most organisms and the formation of mats of sulfur bacteria as mentioned above. Such conditions are, clearly, not conducive to a productive fishery.

In many temperate regions, stratification of the water column is a seasonal phenomenon. When winter comes and stratification is weakened or broken down, oxygen is mixed into bottom waters and hypoxia/anoxia relieved. Recolonization by oxygen-requiring organisms of the bottom can now proceed but, if hypoxia/anoxia reoccurs during the following summer(s), the recolonization process will be interrupted. As long as this pattern of annual (or nearly annual) hypoxia events persists, the benthic fauna community will not be able to reestablish its prehypoxia species and age distribution.

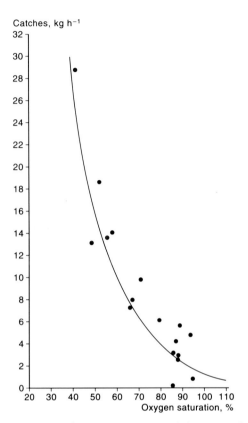

Figure 1.4. Catches (kg trawl h⁻¹) of the Norwegian lobster (*Nephrops norvegicus*) as a function of oxygen concentration in bottom waters. From Bagge [1977].

1.10. Eutrophication of a Marine Coastal Ecosystem: A Danish Case Study

This book deals with the different modes of delivery of nutrients to a marine ecosystem (chapters 2 and 3) and the effects of eutrophication on the various trophic levels occurring within the ecosystem (chapters 4–12). The principles discussed are applicable to most temperate coastal marine systems. However, we have chosen to illustrate many of the concepts addressed with examples from the Kattegat and surrounding bays and fjords (Figure 1.5). This presentation form has been chosen in order to facilitate comparison of the relative importance of various processes in nutrient turnover in the system and to emphasize the importance of obtaining an overall description of nutrient turnover in an ecosystem in order to assess the implications of eutrophication.

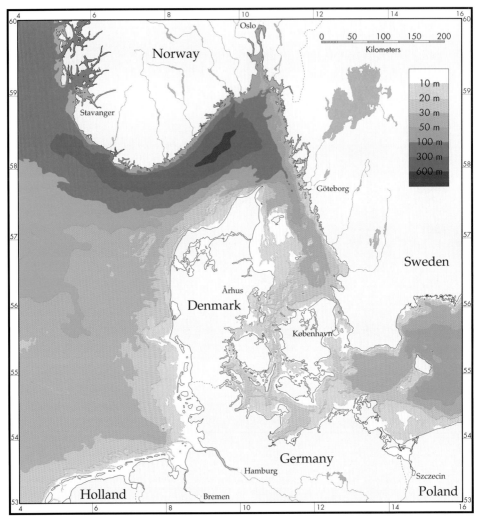

Figure 1.5. Bathymetry of the transition area between the North Sea and the Baltic Sea. (Map design and research: Jonathan Wyss, Topaz Design and Thorkild Aarup.)

The Kattegat represents the transition area separating the Baltic from the Skagerrak and North Sea. From Figure 1.5, it can be seen that this is a relatively shallow water body. In addition, there are a number of bays and inlets along the Danish and Swedish coasts. Water volume and circulation in these inlets is reduced in relation to that occurring in the more open regions of the Kattegat. Thus, nutrients entering these inlets will potentially influence the local ecosystem differently than those entering the more open regions of the Kattegat (see chapters 9 and 10).

The net circulation pattern in the Kattegat is one in which bottom water of relatively high salinity (entering from the Skagerrak in the north) flows toward the Baltic and surface water flows from the Baltic toward the North Sea (see chapter 3). This

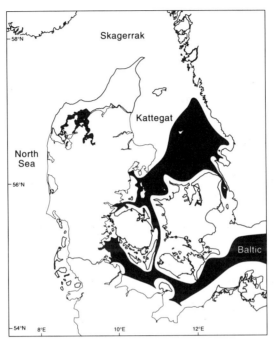

Figure 1.6. Distribution of hypoxia (< 3 mg O_2 l^{-1}) in the Kattegat in 1988 [HELCOM, 1990.]

surface outflow is driven by fresh-water input to the Baltic. Thus, the surface waters in the Kattegat are considerably less saline than the bottom waters.

Hypoxia in the Kattegat's bottom waters began to occur regularly over large areas during the 1980s and is now an annually occurring phenomenon is this region – although the intensity and geographic distribution of this hypoxia varies from year to year depending on water exchange conditions. The most extensive hypoxia event recorded to date occurred in 1988 (Figure 1.6). This recurring hypoxia has affected the local fishery and has caused public outcry. In the aftermath of the realization that cultural eutrophication is the most likely cause of the observed changes in the marine ecosystem of the Kattegat, many questions are being asked concerning the nutrient sources which are most responsible for the observed eutrophication effects and the relative value of different proposed remedial actions.

It is the responsibility of the scientific community to develop the necessary expertise to answer such questions. This book represents our attempt to collate and synthesize the existing expertise with respect to how coastal marine ecosystems respond to eutrophication. It is our hope that this synthesis will help in the answering of some of these questions but also in generally drawing the works of the scientific and the administrative communities closer to one another by providing both a review of "the state of the art" with respect to marine coastal eutrophication and by stimulating ideas for further directions for research.

References

Bagge, O., Norway Lobster (in Danish), *Fisk & Hav (Fish & Sea)*, Danish Institute for Fisheries and Marine Research, 39–44, 1977.

Billen, G., C. Lancelot, and M. Meybeck, N, P, and Si retention along the aquatic continuum from land to ocean, in *Ocean Margin Processes in Global Change*, edited by R. F. C. Mantoura, J.-M. Martin, and R. Wollast, 9, pp. 19–44, John Wiley & Sons, Chichester, 1991.

Cadée, G. C., and J. Hegeman, Seasonal and annual variation in *Phaeocystis pouchetti (Haptophyceae)* in the westernmost inlet of the Wadden sea during the 1973 to 1985 period, *Neth. J. Sea Res.*, 20(1), 29–36, 1986.

Codispoti, L. A., Phosphorus vs. nitrogen limitation of new and export production, in *Productivity in the Ocean: Present and Past*, edited by W. H. Berger, V. S. Smetacek, and G. Wefer, John Wiley & Sons, Chichester, 1989.

Conley, D. J., C. L. Schelske, and E. F. Stoermer, Modification of the biogeochemical cycle of silica with eutrophication, *Mar. Ecol. Prog. Ser.* 101, 179–192, 1993.

Cushing, D. H., A difference in structure between ecosystems in strongly stratified waters and in those that are only weakly stratified, *J. Plank. Res.*, 11(1), 1–13, 1989.

Dugdale, R. C., and J. J. Goering, Uptake of new and regenerated forms of nitrogen in primary productivity, *Limnol. Oceanogr.*, 12, 196–206, 1967.

Edvardsen, B., F. Moy, E. Paasche, Hemolytic activity in extracts of *Chrysochromulina polylepis* grown at different levels of seenite and phosphate, in *Toxic Phytoplankton Blooms in the Sea*, edited by E. Granéli, B. Sundström, L. Edler, and D. M. Anderson, pp 284–289, Elsevier, Amsterdam, 1990.

Goldman, J. C., J. J. McCarthy, D. G. Peavey, Growth rate influence on the chemical-composition of phytoplankton in oceanic waters, *Nature*, 279, 210–215, 1979.

Granéli, E. Paasche, and S. Y. Maestrini, Three years after the *Chrysochromulina polylepis* bloom in Scandinavian waters in 1988: Some conclusions of recent research and monitoring, in *Toxic Phytoplankton Blooms in the Sea*, edited by T. J. Smayda, and Y. Shimizu, 3, pp. 23–32, Elsevier, Amsterdam, 1993.

Gray, J. S., Eutrophication in the sea, in *Marine Eutrophication and Population Dynamics*, edited by G. Colombo, I. Ferrari, V. U. Ceccherelli, and R. Rossi, pp. 3–15, Olsen & Olsen, Fredensborg, Denmark, 1992.

Greve, W., and T. R. Parsons, Photosynthesis and fish production: Hypothetical effects of climate change and pollution, *Helgoländer wiss. Meeresunters.* 30, 666–672, 1977.

Hallegraeff, G. M., A review of harmful algal blooms and their apparent global increase, *Phycologia*, 32(2), 79–99, 1993.

Hansson, S., and L. G. Rudstam, Eutrophication and Baltic fish communities, *Ambio*, 19, 123–125, 1990.

Hecky, R. E., and P. Kilham, Nutrient limitation of phytoplankton in freshwater and marine environments: a review of recent evidence on the effects of enrichment, *Limnol. Oceanogr.*, 33(4(2)), 796–822, 1988.

Heilmann, J. P., K. Richardson, and G. Ærtebjerg, Annual distribution and activity of phytoplankton in the Skagerrak-Kattegat frontal region, *Mar. Ecol. Prog. Ser.*, 112, 213–223, 1994.

HELCOM, Second periodic assessment of the state of the marine environment of the Baltic Sea 1984–1988. Background Document, *Balt. Sea. Envir. Proc.*, 35B, 432 pp., 1990.

Howarth, R.W., Nutrient limitation of net primary production in marine ecosystems, *Ann. Rev. Ecol.*, 19, 89–110, 1988.

Kiørboe, T., Turbulence, phytoplankton cell size, and the structure of pelagic food webs, *Adv. Mar. Biol.*, *29*, 1–72, 1993.

Larsson, U., R. Elmgren, and F. Wulff, Eutrophication and the Baltic Sea: Causes and Consequences. *Ambio*, *14*(1), 9–14, 1985.

Maestrini, S. Y., and E. Granéli, Environmental conditions and ecophysiological mechanisms which led to the 1988 *Chrysochromulina polylepis* bloom: an hypothesis, *Oceanol. Acta* *14*(4), 397–413, 1991.

Mantoura, R. F. C., J. M. Martin, and R. Wollast (Eds.), *Ocean Margin Processes in Global Change*, 469 pp., John Wiley & Sons, Chichester, 1991.

Nixon, S. W., Coastal marine eutrophication: a definition, social causes, and future concerns, *Ophelia*, *41*, 199–220, 1995.

Noe-Nygaard, N., F. Surlyk, and S. Piasecki, Bivalve mass mortality caused by toxic dinoflagellate blooms in a Berrisian-Valanginian lagoon, Bornholm, Denmark, *Palaios*, *2*, 263–273, 1987.

Noe-Nygaard, N., and F. Surlyk, Washover fan and brackish bay sedimentation in the Berrisian-Valanginian of Bornholm, Denmark, *Sediment. 35*, 197–217, 1988.

Piasecki, S., Dinoflagellate cyst stratigraphy of the Lower Cretaceous Jydegård Formation, Bornholm, Denmark, *Bull. Geol. Soc. Denmark*, *32*; 145–161, 1984.

Piasecki, S., Palynological analysis of the organic debris in the Lower Cretaceous Jydegård Formation, Bornholm, Denmark, *Grana*, *25*, 119–129, 1986.

Prakash, A., Coastal organic pollution as a contributing factor to red-tide development, *Rapp. P.-v. Réun. Cons. int. Explor. Mer, 187*, 61–65, 1967.

Richardson, K., and G. Ærtebjerg, Nitrogen, phosphorus, and organic material in the terrestrial and marine environment, in *Report from a Consensus Conference* (in Danish), Danish Ministry of Education and Research, 1991.

Rodhe, W., Crystallization of eutrophication concepts in northern Europe, in *Eutrophication: Causes, Consequences, Correctives*, 50–64 National Academy of Sciences, Washington, DC, 1969.

Rosenberg, R., Eutrophication – the future marine coastal nuisance? *Mar. Poll. Bull. 16*(6), 227–231, 1985.

Rosenberg, R., R. Elmgren, S. Fleischer, P. Jonsson, G. Persson, and H. Dahlin, Marine eutrophication case studies in Sweden, *Ambio*, *19*(3), 102–108.

Smayda, T. J., Novel and nuisance phytoplankton blooms in the sea: evidence for a global epidemic, in *Toxic Marine Phytoplankton*, edited by E. Granéli, B. Sundström, L. Edler, and D. M. Anderson, pp. 20–40, Elsevier, New York, 1990.

Smetacek, V., U. Bathmann, E.-M. Nöthig, and R. Scharek, Coastal eutrophication: causes and consequences, in *Ocean Margin Processes in Global Change*, edited by R. F. C. Mantoura, J.-M. Martin, and R. Wollast, pp. 251–279, John Wiley & Sons, Chichester, 1991.

Smith, S. V., and J. T. Hollibaugh, Carbon-controlled nitrogen cycling in a marine "macrocosm": an ecosystem-scale model for managing cultural eutrophication. *Mar. Ecol. Prog. Ser.*, *52*, 103–109, 1989.

Tatara, K., Utilization of the biological production in eutrophicated sea areas by commercial fisheries and the environmental quality standard for fishing ground. *Mar. Poll. Bull.*, *23*, 315–319, 1991

Vollenweider, R. A., Advances in defining critical loading levels for phosphorus in lake eutrophication. *Mem. Ist. It. Idrobiol.*, *33*, 53–84, 1976.

Vollenweider, R. A., Coastal marine eutrophication: principles and control, in *Marine Coastal Eutrophication*, edited by R. A. Vollenweider, R. Marchetti, and R. Viviani, pp. 1–20, Elsevier, Amsterdam, 1992.

2

Atmospheric Processes

Willem A.H. Asman and Søren E. Larsen

2.1. Introduction

Nitrogen and phosphorus can be limiting factors for algal growth in the marine environment [Dugdale, 1967; Ryther and Dunstan, 1971]. The atmospheric phosphorus input to estuarine and coastal waters is negligible compared to the contribution from other sources. Atmospheric nitrogen input is, however, potentially significant and contributes up to 20–50% of the external nitrogen loading of these waters [Paerl and Fogel, 1994]. For these reasons, only atmospheric nitrogen input is discussed in this chapter.

The most important groups of atmospheric nitrogen compounds that act as nutrients are:

- The NH_x group: gaseous ammonia (NH_3), and ammonium in particles (NH_4^+ aerosol).
- The NO_y group: the gaseous compounds dinitrogen pentoxide (N_2O_5), HNO_4 and nitric oxide (NO), nitrogen dioxide (NO_2), nitric acid (HNO_3), nitrous acid (HNO_2), peroxy acetyl nitrate (PAN) nitrate in particles (NO_3^- aerosol) and nitrate radicals (NO_3).
- Organic nitrogen compounds. The atmospheric concentrations of these compounds are very low and their sources are not well known. For these reasons organic nitrogen compounds are not treated in this chapter, apart from PAN that is a relatively well-known reaction product.

The combination of NH_3 and NH_4^+ is called NH_x; that of NO and NO_2 is called NO_x. The dominant NO_y compounds at sea are NO_2, HNO_3, PAN, and NO_3^- aerosol. Individual aerosol particles consist of numerous compounds. The particles can differ markedly in size (0.1–5 μm) and their chemical compositions are a function of their size.

Eutrophication in Coastal Marine Ecosystems
Coastal and Estuarine Studies, Volume 52, Pages 21–50
Copyright 1996 by the American Geophysical Union

The atmospheric nitrogen input is caused by two processes: wet deposition, which is the removal from the atmosphere by precipitation and by dry deposition, which is the removal by atmospheric turbulence at the sea or land surface. There are two important differences between wet and dry deposition. Removal by wet deposition takes place over the whole atmospheric layer, where snow and raindrops are found. Removal by dry deposition occurs only very near the sea or land surface. Wet deposition occurs only when precipitation occurs, which in Western Europe is approximately 5–10% of the time. Dry deposition is constantly occurring, even when it rains.

Atmospheric deposition to the sea is not easily measured. Although concentrations in precipitation can be measured at sea, it is not easy to determine the amount of precipitation taking place over the sea. This is not easy because the collection efficiency is heavily influenced by the disturbed wind field around the ship or the platform. Dry deposition is even more difficult to measure and it is not possible to do so routinely. For these reasons, it is difficult to estimate the atmospheric nitrogen input to the sea from measurements and, therefore, atmospheric transport models are often used to calculate the input [van Jaarsveld, 1992]. In atmospheric transport models, it is necessary to describe all processes influencing the concentration of all involved components in an air parcel on its way to where deposition occurs. Theses processes include: emissions, transport by the wind, mixing, reaction and deposition. Deposition processes should not only be known for sea areas but also for land areas because the deposition to land areas determines how much is left for deposition to the sea. The change in mass of a compound in an air parcel during transport can be described by the following simplified differential equation:

rate in change of mass = emission rate + formation rate by reaction
– removal rate by reaction – wet deposition rate – dry deposition rate
+ rate of change due to other exchange mechanisms   (1)

The formation and removal rate of one compound depends on the concentration of other compounds and, in this way, the differential equations of all compounds are intertwined. This complicates the integration of the differential equations in time/space, which is needed to find the concentration/deposition at sea. The annual deposition in a sea area is computed by following the air parcels on their way to the area during one year and calculating the concentration and deposition with (1). The main sources of the nitrogen input to the estuaries and the coastal areas are land-based and a large fraction of this input does not originate from sources in coastal areas but rather from sources more than a few hundred kilometers away. For this reason, it is necessary to consider a rather large area in model calculations.

This chapter discusses the various atmospheric processes that are part of atmospheric transport models with the emphasis on the deposition processes. The detailed setup of such models is not discussed. However, at the end of this paper an example of the results of an atmospheric transport model, 'ACDEP', [Asman et al. 1994a] is presented for the Kattegat. Estimates for other marine regions are also presented.

2.2. Emissions

Ammonia

The predominant atmospheric NH_3 sources are livestock wastes, with somewhat smaller contributions from fertilizer application and production [Buijsman et al., 1987]. In Figure 2.1, the geographical distribution of the NH_3 emission density in Europe is presented. The total emission in Europe (excluding the former USSR) in 1989 was 4534 ktonne N yr^{-1}. Cattle, pigs and the application of fertilizers are the most important sources and contribute, 53, 18 and 17% of the NH_3 emission, respectively. The emission density is relatively high in Denmark, though lower than in The Netherlands and Belgium. Since 1950, European NH_3 emissions have doubled [Asman et al., 1988]. The seasonal variation in the NH_3 emission rate is poorly known. It is guessed that the rate is highest in spring and summer [Asman, 1992].

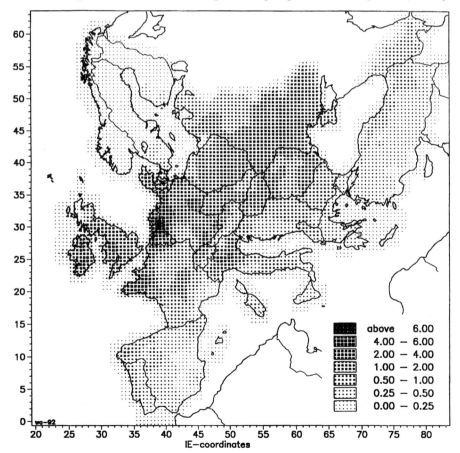

Figure 2.1. Emission density of NH_3 in Europe (tonne N km^{-2} yr^{-1}) without Russia.

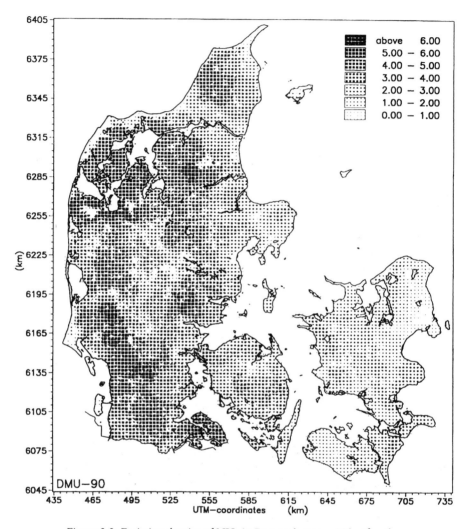

Figure 2.2. Emission density of NH_3 in Denmark (tonne N km^{-2} yr^{-1}).

The NH_3 emission rate is generally higher during daytime than during nighttime, which is the result of a higher temperature and wind speed during daytime and the fact that manure is mainly spread during daytime. NH_4^+ aerosol is not emitted to the air. This means that all atmospheric NH_4^+ originates from gaseous NH_3.

The NH_3 emission in Denmark is 109 ktonne N yr^{-1} [Asman et al., 1993]. Cattle and pigs are the most important sources and contribute 46 and 37%, respectively, to the total NH_3 emissions. Figure 2.2 shows the geographical distribution of the NH_3 emission density in Denmark. The NH_3 emission density varies on a regional basis, both on European as well as Danish scales. The sea may also act as a minor source of NH_3 emission [Quinn et al., 1988a,b, and 1990; Asman et al., 1994b].

Nitrogen Oxides

Anthropogenic emission of NO_x is dominated by the contribution of NO. Most of the NO_x emission is generated during the combustion of fossil fuels and originates mainly from N_2 and O_2 present in the air needed for the combustion. The total emission of NO_x in Europe, excluding the former USSR, is 5094 ktonne N yr^{-1}, of which 2427 is from stationary sources and 2667 from mobile sources [Pacyna et al., 1991]. In Western European countries, the contribution of mobile sources (road traffic, internal navigation) is about 60% of the total. For Eastern European countries this was in 1991 about 40%, but it is certainly increasing. The geographical distribution of the NO_x emission density in Europe is shown in Figure 2.3. The NO_x emission density is highest in the most densely populated areas of Europe and is, for that reason, somewhat lower in Denmark. The European NO_x emission has increased by a factor 3 since 1950 [Pacyna et al., 1991]. The NO_x emission rate is

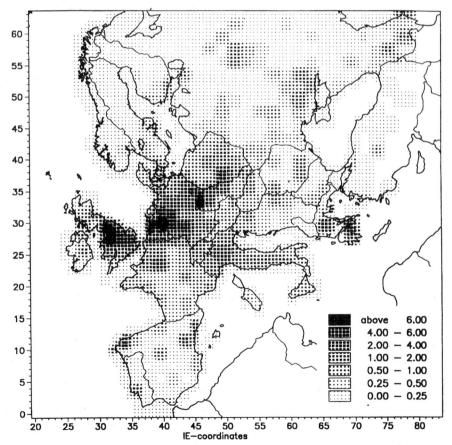

Figure 2.3. Emission density of NO_x in Europe (tonne N km^{-2} yr^{-1}).

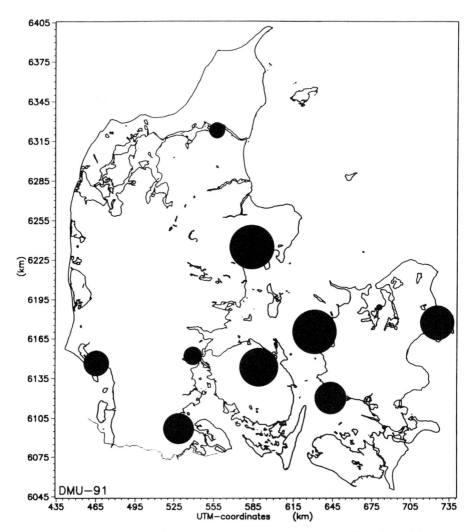

Figure 2.4. Emissions of NO$_x$ from large point sources in Denmark. The radius of the circle indicates the relative source strength.

highest during daytime during rush hours and during the winter heating season. All other NO$_y$ compounds are not emitted to the air, but are reaction products of NO$_x$.

The NO$_x$ emission in Denmark is 89 ktonne N yr^{-1}. The most important sources are large point sources (power plants etc.: 44%) and road traffic (34%) [Asman et al., 1993]. The geographical distribution of the emissions is divided into large point sources (mainly power plants; Figure 2.4) and area sources (Figure 2.5).

NO$_x$ emissions in Europe are about as large as NH$_3$ emissions. The uncertainty in both nitrogen emissions is at least 30–40%. The emission of other compounds that

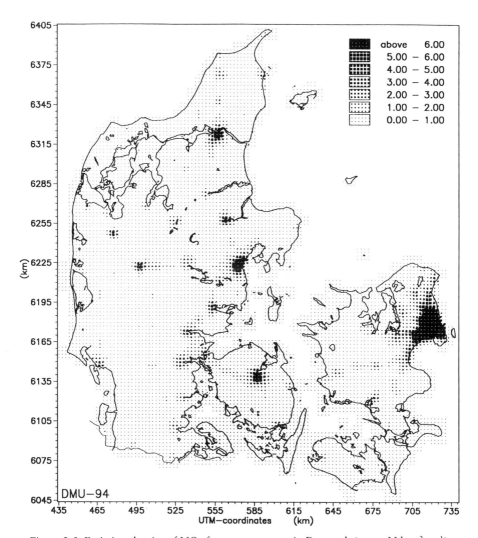

Figure 2.5. Emission density of NO_x from area sources in Denmark (tonne N km^{-2} yr^{-1}).

play a role in the atmospheric chemistry of nitrogen compounds like SO_2 and hydro-carbons, is not discussed here because their influence is indirect. In atmospheric transport models, gridded emission data are used. In such models, it is assumed that the emission is homogeneously distributed within each grid element. In Europe and North America, grid sizes of the order of 100–150 km are used. The use of such sizes can cause problems in coastal areas, where the emission often only occurs on the land area within the grid element. The dry deposition at sea will in such cases be overestimated in the model because the air concentration is too high due to the too high emission density at sea. To avoid this problem an emission grid size of 15 km was used in the ACDEP model to calculate the deposition to the Kattegat.

2.3. Reactions

Compounds are chemically converted through atmospheric reactions. Each compound has its own specific physical and chemical characteristics, and it is therefore removed from the atmosphere at a different rate than that of other compounds. With respect to deposition, reactions are very important. For example, NH_3 is removed relatively quickly from the atmosphere by dry deposition, as compared to its reaction product, NH_4^+ aerosol. This, combined with the relatively fast reaction from NH_3 to NH_4^+ aerosol, results in a transport of NH_x over longer distances than would be the case if the reaction were slow.

Under Western European conditions, most NH_3 will react with acid aerosol, e.g. aerosol containing sulphuric acid (H_2SO_4). Although this is a fast reaction in the laboratory [Robbins and Cadle, 1958; Baldwin and Golden, 1979; Huntzicker et al., 1980; McMurry et al., 1983], it is found that the overall conversion rate of NH_3 in the atmosphere is much lower, namely about 30% h^{-1} [Erisman et al., 1988; Asman and van Jaarsveld, 1992]. This cannot yet be explained. However, it may be that all acid aerosol near the earth's surface has reacted with NH_3 and that NH_3, therefore, has to be transported higher up in the atmosphere before it can react. This transport takes time and this could be one of the reasons why the conversion rate from NH_3 to NH_4^+ aerosol in the atmosphere is not as high as in the laboratory. Once bound to SO_4^{2-} in particles as NH_4^+, NH_3 is not released again. A minor part of NH_3 reacts with HNO_3 and HCl to form NH_4NO_3 and NH_4Cl-containing particles. Reactions:

$$NH_3 + HNO_3 \rightleftarrows NH_4NO_3 \qquad (2)$$
$$NH_3 + HCl \rightleftarrows NH_4Cl \qquad (3)$$

These reactions can proceed in both ways, which means that NH_3 (and HNO_3 and HCl) not only can be consumed, but can be produced [Stelson et al., 1979; Stelson and Seinfeld, 1982 a-c; Pio and Harrison, 1987; Allen et al., 1989; Ottley et al., 1992; Harrison and Kitto, 1992; Mozurkewich, 1993]. In the USA, these reactions are more important than in Western Europe, because the ratio of NO_x to SO_2 emission is larger. Over the sea, the NH_3 concentration is rather low and for that reason there is often not enough NH_3 to form NH_4NO_3 and NH_4Cl-containing particles. The equilibrium constants of these reactions are temperature and relative humidity-dependent.

The reactions of NO_y are much more complex because many other compounds are involved (Figure 2.6). Some of the reactions are influenced by sunlight and very reactive compounds, radicals, are formed. Because they are so reactive, they occur only in very low concentrations, but their role in atmospheric chemistry during daylight hours is very important. Examples of these radicals are RO_2, $RCHO$ and $RO(O)_2$, all organic forms, and the inorganic ones OH and NO_3 (these radicals are not in ion-form; R denotes a saturated hydrocarbon chain; NO_3 denotes NO_3 radicals). In addition, other reactions are also directly influenced by sunlight, which is indicated in Figure 2.6 by '$h\upsilon$'. NO_x is emitted mainly to the atmosphere in the form of NO. NO is rapidly oxidized to NO_2 by reaction with ozone (O_3) and RO_2.

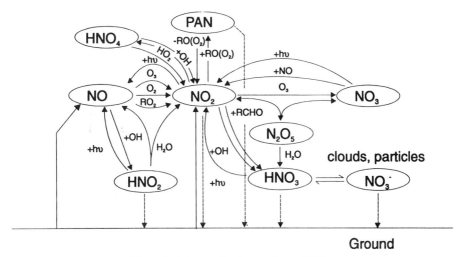

Figure 2.6. Atmospheric reactions of NO_y.

At the same time, HNO_2 is also formed. NO_2 can react to PAN, HNO_3, NO_3 radicals, and N_2O_5. N_2O_5 is a gaseous compound that is stable only at night, as it is photolyzed by sunlight. HNO_3 or N_2O_5 can react further to nitrate (NO_3^-)-containing aerosols or be taken up by cloud- and raindrops, which also leads to NO_3^- formation. The atmospheric chemistry used in the ACDEP model is described by Gery et al. [1989] and Hertel et al. [1993].

The fact that a compound has a high concentration does not imply that the compound is an important contributor to deposition, for this will not only depend on the concentration, but also on the efficiency of the removal mechanism. In fact, the opposite can occur: namely that the concentrations of some compounds are high because they are not removed at a high rate.

2.4. Meteorological Aspects

Amount of Precipitation

The atmospheric removal mechanism that contributes most to the nitrogen deposition in marine areas is wet deposition. The amount of wet deposition generally increases with the amount of precipitation. It is therefore important to know the amount of precipitation at sea. The amount of precipitation is very difficult to measure on ships and platforms because the collection efficiency of the rain gauge is not very well known as the air flow around the ship disturbs the collection. It is generally assumed that the amount of precipitation at sea is lower than in coastal regions. Meteorological weather prediction models give a mean annual amount of precipitation for the North Sea of 550 mm [Krell and Roeckner, 1988; Petersen et al., 1988]. Precipitation measurements on oil and gas platforms in the North Sea are of the order of 400 mm yr^{-1} [Rendell et al., 1993]. The seasonal variation in the rainfall at sea can be different from that measured on land. This is caused by the fact that the temperature of the sea water is different from the temperature on land. During the summer, the temperature on land near the surface can become rather high compared to the air aloft, which can lead to convective precipitation on land. During autumn, the temperature of sea water is often higher than the air temperature, leading to more convective precipitation at sea in autumn.

The amount of precipitation over the Kattegat was determined from data at 21 coastal sites in Denmark and Sweden and on several islands [Allerup et al., 1992]. This amount was compared with data from 20 stations on the Jutland peninsula. Over the four-year period 1988–1991 precipitation on the Kattegat was 84% of that found on Jutland. The ratio between the monthly amounts of precipitation on Jutland and the Kattegat varies from 0.6 to 1.0. The highest ratio seems to occur in the late autumn and early winter. This confirms the trend discussed above. Model calculations [Asman et al., 1994a] showed that it was impossible to reproduce the wet deposition of nitrogen compounds measured on the island of Anholt in the Kattegat with the amount of precipitation measured on Jutland. This illustrates, that it is important to use the correct amount of precipitation for sea areas in atmospheric transport models and not, as is often done, values from land areas.

Turbulence and Transport in the Atmosphere

The lower surface roughness height for momentum at sea causes a higher wind speed, but less turbulence at sea than on land [Garratt, 1990; Gryning, 1993]. The reduced turbulence will for many compounds lead to lower dry deposition velocities at sea than on land (see section on dry deposition). The atmosphere at sea tends often to be stable in winter and spring and unstable in summer and autumn. This is caused by differences in temperature between air and sea. Under stable conditions the wind speed is reduced, which leads to reduced dry deposition velocities, whereas they are increased under unstable conditions.

Transport in the atmosphere is usually described by wind fields, which are based on meteorological observations. In Eulerian models this information is used directly. In Lagrangian models this information is used to calculate trajectories, i.e. the path an air parcel travels as a function of time under influence of the varying wind fields. Usually, back-trajectories are used in calculations, i.e. the traject to a point, a so-called receptor point, is followed. Wind direction and wind speed are functions of height. This means that air at ground level has an origin different from air aloft.

Atmospheric turbulence causes air to be exchanged between different heights. In atmospheric transport models this exchange is often modeled by using K-theory [Stull, 1988]:

$$\frac{\partial c}{\partial t} = \frac{\partial}{\partial z}\left(K_z \frac{\partial c}{\partial z}\right)$$

$$(4)$$

Here, the change in concentration is dependent on changing concentrations between vertical layers in the model.

Solubility of Gases in Water

Solubility in water is a very important property of gases. It determines to a large extent the rate at which gases are removed by dry and wet deposition. The solubility is expressed in the Henry's law coefficient (H_A expressed in mol l^{-1} atm^{-1}) which for species A is defined here as

$$H_A = \frac{[A.H_2O]}{p_A} \qquad (5)$$

where $[A.H_2O]$ is the concentration of the dissolved gas in water (mol l^{-1}) and p_A is the pressure of the gas in air (atm). The larger a Henry's law coefficient, the more soluble the gas is. It should be mentioned here that there exist different definitions of Henry's law coefficient, which can be confusing.

For compounds that dissociate, like acids and bases, even more of the compound can dissolve in water. This means that the amount dissolved at a certain concentration in air is also a function of the pH of the water. The pH of cloud and precipitation water is usually between 4 and 5, whereas the pH of sea water is around 8. The Henry's law coefficients for nitrogen compounds are shown in Tables 2.1 and 2.2. They show that NO, NO_2 and PAN are poorly soluble in water. NO and NO_2

TABLE 2.1. Henry's law coefficients H at 25°C (mol l^{-1} atm^{-1}) and their temperature coefficients, $-\Delta H/R$ (°K).

Compound	H	$-\Delta H/R$	Reference
NH_3	$5.60 \times 10^{+1}$	4092	Dasgupta and Dong [1986]
NO	1.93×10^{-3}	1479	Schwartz and White [1981]
NO_2	1.20×10^{-2}	1965	Schwartz and White [1981]
HNO_2	$4.90 \times 10^{+1}$	4781	Schwartz and White [1981]
HNO_3	$2.10 \times 10^{+5}$	8706	Schwartz and White [1981]
NO_3 radical	very large		
N_2O_5*	very large		
PAN	2.80	6513	Kames et al. [1991]

* Reacts with water to HNO_3 (see HNO_3).

TABLE 2.2. Dissociation: constants K at 25°C and their temperature coefficients, $-\Delta H/R$ (°K). Concentrations are in mol l^{-1}.

Compound	K	$-\Delta H/R$	Reference
NH_3 *	1.78×10^{-5}	-429	Bates and Pinching [1950]
HNO_2	5.01×10^{-4}	-1258	Schwartz and White [1981]
HNO_3 **	$1.55 \times 10^{+1}$	0	Schwartz and White [1981]

* $NH_3.H_2O \leftrightarrow NH_4^+ + OH^-$
** temperature dependence included in Henry's law coefficient (Table 2.1).

react in water, but not very fast compared to the dissociation reaction mentioned before, to form NO_2^- (nitrite) and NO_3^- (nitrate) [Schwartz and White, 1981]. PAN is hydrolyzed in water, which proceeds at a reasonable speed [Kames et al., 1991]. The reaction products are NO_2^- and NO_3^-. The other gases are soluble in water.

The temperature dependence of Henry's law coefficient is described by

$$H(T) = H(298.25) \exp\left[\frac{-\Delta H}{R} \left(\frac{1}{T} - \frac{1}{298.15} \right) \right] \tag{6}$$

where T is the temperature in K. This equation is also valid for the dissociation constants, but then one should read K instead of H. For sea water, it is necessary to take into account that the ionic strength is so large, that it influences Henry's law coefficients and the dissociation constants [Millero and Schreiber, 1982].

2.6. Wet Deposition

Wet deposition refers both to the process of wet deposition and to the amount deposited by this process. Compounds can be removed by different wet deposition processes. There exist removal processes within clouds (in-cloud scavenging) and removal processes below-cloud base (below-cloud scavenging), where compounds are removed by falling raindrops and snowflakes.

The atmosphere contains aerosol particles which consist partly of hygroscopic substances. When the relative humidity is more than 40%, aerosols contain at least 30% water by weight. When the relative humidity increases, more water vapor condenses onto the aerosols and cloud droplets are formed. Most of the NH_4^+ and NO_3^- is found in those aerosols, which can act as condensation nuclei. Most cloud droplets will evaporate again and it is estimated that the cycle, aerosol → cloud droplet → aerosol, is repeated at least ten times before the aerosol is removed by precipitation formed from cloud droplets. The process of removal of aerosols due to their role as condensation nuclei is the most important process for in-cloud removal of particles. Removal of aerosols by below-cloud scavenging is not very efficient.

The flux of gases into droplets can be described by the following equation [Peters, 1983]:

$$F_g = \frac{f_g D_g}{r} (c_g - c_w^*) \tag{7}$$

where

F_g = flux into the drop (mol m^{-2} s^{-1})

f_g = ventilation coefficient (dimensionless), which is the ratio of uptake of a drop moving relative to the air and a non-moving drop. The ventilation coefficient is needed to describe the effect of enhanced uptake due to internal circulation in the drop, which increases with fall speed and therefore also with drop size (the fall speed increases with drop size).

D_g = diffusivity of the gas (m^2 s^{-1})
r = radius of the drop (m)
c_g = gas phase concentration of the gas (mol m^{-3})
c_w* = the theoretical gas phase concentration, which would be in equilibrium with
      the concentration of dissolved gas at the gas droplet interface (mol m^{-3}).

The value of c_w* is found from the concentration in the liquid phase:

$$c_w{}^* = \frac{c_w}{HRT} \tag{8}$$

c_w = concentration of the dissolved gas in the liquid phase (mol m^{-3})
H = Henry's law coefficient (mol l^{-1} atm^{-1})
R = gas constant (atm l^{-1} mol^{-1} K^{-1})
T = temperature (K).

The flux rate will thus depend on the concentration difference between gases and liquid phases. Some time fóllowing exposure of a drop to gas concentrations, equilibrium will be reached between the phases. For the same gas, it will take more time to reach equilibrium for large drops (rain drops, radius 100–2500 μm) than for small drops (cloud droplets, radius about 10 μm). The size and the lifetime of cloud drops is such that they will reach equilibrium with the surrounding (interstitial) air within a few seconds. For moderately soluble gases, the concentration in the interstitial air will be almost the same as the concentration in the air before the cloud was formed because only a small fraction of the gas dissolves into the droplets. The concentration below the cloud will usually not be much different from that in the cloud, i.e. there exists near equilibrium between rain drops formed out of cloud droplets and the surrounding air below the cloud. For that reason, they will not absorb much gas below the cloud.

In the case of highly soluble gases, the largest fraction of the gas will dissolve into the droplets and the concentration in the interstitial air will become very low compared to the situation before cloud formation. This means that these drops will be far from equilibrium with the air below the cloud. The maximum uptake rate of the raindrops is much lower than of the cloud droplets and their lifetime is much shorter. For these reasons, the drops will not get saturated with the gas below cloud base and they are, therefore, not saturated when they are collected at the ground. This will be the situation for NH_3, HNO_3, N_2O_5 [Asman, 1994]. This situation can cause artifacts, because gas can be taken up in the rain in the sampler after collection.

The cloud base is often a few hundred meters high, whereas clouds usually are much thicker. Therefore, the volume exposed to below-cloud scavenging is normally small compared to the volume exposed to in-cloud scavenging and, as a result, so is the relative contribution of below-cloud scavenging to the concentration in precipitation.

When NH_3 is absorbed by cloud and rain drops, which are usually acidic, it will react with H^+ to form NH_4^+. This NH_4^+ cannot be distinguished from the NH_4^+ originating from NH_4^+-containing aerosol. The same holds for NO_3^- in precipitation, which originates from the scavenging of HNO_3, N_2O_5 and NO_3^--containing aerosol. It is also nearly impossible to find the contribution of below-cloud and in-cloud scavenging from measurements. Only models can produce this kind of information.

Wet deposition is influenced by the air flow in and around the clouds, cloud physics and cloud chemistry. A simple way to express the removal rate of airborne compounds (both gases and aerosols) is the scavenging coefficient λ, which is the fraction of the airborne concentration removed per unit of time. The airborne concentration c_g can then be described by

$$c_g = c_{g,0} \exp(-\lambda t) \tag{9}$$

where

$c_{g,0}$ = concentration at $t = 0$
λ = scavenging coefficient (s^{-1})
t = time (s).

Scavenging coefficients depend on rainfall rate, which is the result of a series of processes. Scavenging coefficients can be used to describe removal by in-cloud processes and below-cloud processes separately or the removal by both processes. For gases, in- and below-cloud scavenging coefficients are functions of Henry's law coefficient if equilibrium is established between the air and the drop. If no equilibrium is reached, which is the case for highly soluble gases below the cloud, the coefficient will depend on the diffusivity of the gas and on the ventilation coefficient, which is a function of the radius [Levine and Schwartz, 1982; Asman, 1994].

Scavenging coefficients cannot be used if compounds are involved in complicated reactions in the clouds. This is, however, not the case for NH_x and NO_y. In- and below cloud scavenging coefficients for a rainfall rate of 1 mm h^{-1} are shown in Table 2.3. They are upper estimates. Usually, the coefficients will be smaller because precipitation occurs only in part of the area of interest and mixing height can be much higher. A more realistic number would be about $1/10-1/3$ of the results mentioned in Table 2.3. Most of the NH_3, NH_4^+ aerosol, HNO_3, NO_3 radical, N_2O_5, and NO_3^- aerosol will be removed after one hour of continuous precipitation. NO, NO_2 and PAN are removed at a very low rate by precipitation. When it occurs, wet deposition is a much more efficient removal process than dry deposition (see next section). However, it rains only 5–10% of the time in north-western Europe. This means that wet deposition averaged over a longer period, e.g. a year, is less efficient than suggested by Table 2.3. It can, therefore, be of the same order as the less efficient dry deposition process, because this last process occurs all the time.

2.7. Dry Deposition

Dry deposition refers both to the dry deposition process as well as to the amount deposited by this process. Dry deposition is the transport of airborne compounds to the land and sea surface by atmospheric turbulence. Atmospheric turbulence can be produced either by mechanical processes, i.e. wind which strikes over a rough surface, or by a higher temperature of the surface than of the air, i.e. wind which

TABLE 2.3. Scavenging coefficients (upper estimates) of nitrogen compounds at a rainfall rate of 1 mm h^{-1}. For in-cloud scavenging, a mixing height of 1 km is assumed [Asman et al., 1994a].

Compound	Scavenging coefficient λ (s^{-1})	
	in-cloud	below-cloud
NH$_3$	1.4×10^{-3}	9.5×10^{-5}
NH$_4^+$ aerosol	1.3×10^{-3}	5.0×10^{-6}
NO	1.9×10^{-11}	≈ 0 *
NO$_2$	1.4×10^{-10}	≈ 0 *
HNO$_2$	1.5×10^{-5}	≈ 0 *
HNO$_3$	1.4×10^{-3}	6.2×10^{-5}
NO$_3$ radical	1.4×10^{-3}	6.0×10^{-5}
N$_2$O$_5$	1.4×10^{-3}	4.0×10^{-5}
PAN	1.3×10^{-7}	≈ 0 *
NO$_3^-$ aerosol	1.3×10^{-3}	5.0×10^{-6}

* Almost no uptake will take place.

strikes over the surface. Atmospheric turbulence is produced by two types of processes: a mechanical process, in which the wind is altered by a rough surface; and a convective process, when the sea surface is warmer than the air. Even though dry deposition is calculated with an atmospheric transport model for sea areas only, it is necessary to take dry deposition on land into account. The reason for this is that dry deposition is a sink and partly determines the amount left in the air parcel that can be deposited at sea. For this reason, roughness height and dry deposition on land are discussed here as well. First, surface roughness will be discussed because it is implicated in atmospheric turbulence processes.

Surface Roughness

Table 2.4 show the surface roughness for different surfaces [Arya, 1988]. It appears that sea surfaces are very flat, i.e. that they do not cause as much turbulence as land surfaces.

TABLE 2.4. Surface roughness for momentum for different surfaces.

Surface	Surface roughness (m)
Ice, mud flats	10^{-5}
Sea, large expanses of water	10^{-4}–10^{-3}
Snow (on farmland)	2×10^{-3}
Grass (cut, uncut)	8×10^{-3}–2×10^{-2}
Farmland (summer)	2×10^{-2}–1×10^{-1}
Forest	1
Towns (outskirts)	0.4
Centres of towns (small, large)	0.6, 2

The surface roughness for momentum for the sea is, contrary to almost all land surfaces, not constant, but is a function of the wind speed. The reason for this is that
the surface changes because waves are formed. The mechanically produced roughness length for momentum, z_{0m}, is given by the following equation [Lindfors et al.
1991]:

$$z_{0m} = \frac{0.13\nu}{u_*} + \frac{0.0144u_*^2}{g} \tag{10}$$

where

z_{0m} = surface roughness length for momentum (m)
ν = kinematic viscosity of the air (m^2 s^{-1})
u_* = friction velocity (m s^{-1})
g = gravitation (m s^{-2}).

The first term in equation 10 describes a smooth surface. The second term describes
rough conditions. Up to a wind speed of about 3 m s^{-1} the first term dominates and
the surface roughness decreases with wind speed. At larger wind speeds, waves are
formed and the surface roughness increases with wind speed. The surface roughness for momentum cannot be measured directly, but is calculated from measured
vertical wind profiles.

Aerodynamic Resistance

The resistance to transport in the air from a reference height (z_r) to the height z_{0m} is
the so-called aerodynamic resistance. This resistance describes the transport of all
gases and particles to a thin laminar boundary layer (0.1–1 mm) just above the surface. Only for particles with radius larger than 5 μm is the transport in the atmosphere noticeably influenced by gravitation. The aerodynamic resistance r_a (s m^{-1})
at reference height z_r (m) for all gases and particles is given by [Arya, 1988]:

$$r_a = \frac{1}{\kappa u_*}\left[\ln\left(\frac{z_r}{z_{0m}}\right) - \Psi_h\left(\frac{z_r}{L}\right) + \Psi_h\left(\frac{z_{0m}}{L}\right)\right] \tag{11}$$

where

κ = Von Karman's constant ≈ 0.4
L = Monin-Obukhov length (m); This is a function of the ratio of mechanically to
     thermally produced turbulence.

The function Ψ_h is a correction for atmospheric stability.

For stable and conditions ($L > 0$): $\Psi_h = -5z_r/L$
For unstable conditions ($L < 0$): $\Psi_h = 2\ln((1 + x^2)/2)$, with $x = (1 - 15z_r/L)^{1/4}$
For neutral conditions: $\Psi_h = 0$.

Laminar Boundary Layer Resistance

Gases

The laminar boundary layer resistance r_b for gases (s m^{-1}) is defined by

$$r_b = \frac{1}{\kappa u_*} \ln\left(\frac{z_{0m}}{z_{0c}} \right) \tag{12}$$

where z_{0c} is the surface roughness for concentration of gases at sea (m) and is given is for smooth conditions (Reynolds number $Re = z_{0m}u_*/\nu < 0.15$) by

$$z_{0c} = 30 \ (\nu/u_*) \ \exp(-13.6 \ \kappa \ Sc^{2/3}) \tag{13}$$

and for rough conditions ($Re \geq 0.15$):

$$z_{0c} = 20 \ z_{0m} \ \exp(-7.3 \ \kappa \ Re^{1/4} \ Sc^{1/2}) \tag{14}$$

where

Sc = Schmidt number = ν/D_g, where D_g = diffusivity of the gas (m^2 s^{-1}).

The value of r_b for gases at sea can then be found from (12), (11), (13) and (14).

For land surfaces the ratio z_{0m}/z_{0c} of gases with Sc around 0.6 to 0.8 is approximately independent of wind speed and has the value of 7 to 12 for grassy or tree-covered surfaces, but a value of 2 to 3 for tall trees [Brutsaert, 1991].

This laminar boundary resistance is a somewhat artificial resistance. It is a kind of correction factor, which is needed to find the atmospheric resistance for transport of compounds from the atmospheric resistance for transport of momentum, which can be determined relatively easily.

Particles

Particles contain a substantial fraction of the atmospheric nitrogen compounds. There are several mechanisms by which particles can cross the laminar boundary layer but none of these mechanisms is very efficient. Particles have some properties which cause their laminar boundary layer resistance to be much higher than for gases for both sea and land surfaces. The diffusivity of particles is much lower than for gases. This means that particles are not able to cross the laminar boundary layer efficiently by diffusion. Only very small particles with radius < 0.1 μm, which do not contribute much to the total aerosol mass, have reasonable diffusivities. For particles with a radius larger than about 1 μm, transport is more efficient because then interception and impaction are also important. But a large fraction of the particles, including most nitrogen-containing particles, has a radius between 0.1 and 1 μm and is not easily removed.

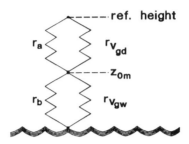

Figure 2.7. Resistance model for dry deposition of particles to the sea.

Wind tunnel measurements have shown that the dry deposition velocities of these particles are low and that they are limited by the transport across the thin laminar boundary layer [Möller and Schumann, 1970; Sehmel and Sutter, 1974]. Field measurements, which are very uncertain, sometimes show much higher dry deposition velocities [Sievering, 1981, 1984; Smith et al., 1991]. These differences were attributed to processes which were not taken into account in the wind tunnel studies, but they could also result from uncertainties in the measurements. Slinn [1983] gives an overview of all processes involved in the air-to-sea transfer of particles. Williams [1982] postulated that sea spray can enhance the transport through the laminar sublayer up to three orders of magnitude by disrupting the layer. Recent experiments in an air-sea exchange tunnel indicate, however, that this effect is negligible [Larsen et al., 1994, 1995]. For this reason, the model of [Slinn and Slinn, 1980; Figure 2.7] was used to calculate the laminar boundary layer resistance r_b for (sea) water:

$$r_b = \frac{\kappa u_{10}}{u_*^2 (Sc^{-1/2} + 10^{-3/St})} \tag{15}$$

where

u_{10} = wind speed at 10 m height (m s^{-1})
Sc = Schmidt number particle = ν/D_p
D_p = Diffusivity particle (m^2 s^{-1})
St = Stokes number = $u_*^2 v_{gw}/(g\nu)$, with v_{gw} = gravitational settling velocity of the humidified particle (m s^{-1}) and
g = gravitation (m s^{-2}).

D_p and v_{gw} can be found in Hinds [1982]. Parallel to the aerodynamic and the laminar boundary layer resistance, there is a resistance due to gravitational settling, which is important only for larger particles. Due to the high relative humidity in the laminar boundary layer, the particles will get humidified which causes them to grow [Fitzgerald, 1975]. This increases their settling velocity and, therefore, their dry deposition velocity. Figure 2.8 shows the results of the model of Slinn and Slinn [1980] for the dry deposition velocity as a function of wind speed and particle size. A minimum in the dry deposition velocity is observed for radii between 0.1 and 1 μm. A substantial fraction of the nitrogen-containing particles is of this size.

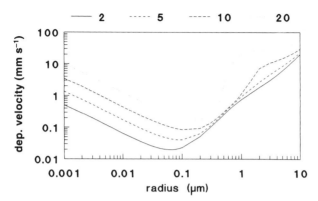

Figure 2.8. Dry deposition velocity of particles at sea as a function of their radius and the wind speed.

For land surfaces r_b for particles is not well known [Davidson and Wu, 1990], but is thought to be rather high, of the order of 500 s m^{-1}. It depends on the particle size and the surface structure, e.g. on the presence of hairs on leaves etc. In general, dry deposition velocity of particles on land shows also a minimum for radii between 0.1 and 1 μm. The value of r_b for land surfaces will be larger than for the sea because vegetation is rougher, which enhances the possibility of capture.

Surface Resistance

Gases

Liss [1983] gives an overview of processes involved in air-to-sea transfer of gases. At the sea, the surface resistance r_c is negligible for highly soluble gases like NH$_3$, HNO$_3$ and N$_2$O$_5$ [Asman et al., 1994c]. For these gases, the resistance to transport to the water is therefore entirely determined by the resistance to transport in the air $(r_a + r_b)$.

In the case of poorly soluble gases the reactivity in the water phase is important. If a gas reacts fast, then the concentration of the dissolved gas in the upper, laminar layer of the water decreases. This enables more of the gas to get dissolved. In this case, the resistance to uptake also becomes lower, but usually not as low as for the highly soluble gases. It mainly depends on the reaction rate. The dry deposition velocity is to a large extent determined by the surface resistance for this type of gases. For that reason, r_c for these gases does not vary much with wind speed. This is partly the case for PAN.

If no reaction or only a slow reaction takes place, than the resistance r_c becomes very large and dominates. This is the case for NO and NO$_2$.

The surface resistance for the uptake of gases by land surfaces depends on the properties of the gas and the surface. Highly soluble and reactive gases like NH$_3$, HNO$_3$

and N_2O_5 will react with all surfaces or will dissolve in water layers on vegetation and their surface resistance will be low. For less soluble gases, transport through the stomata and subsequent absorption in the leaves is important. Stomata, however, are only open during daytime. Consequently a diurnal variation in the surface resistance and dry deposition velocity will occur. This is the case for e.g. NO_2. Such mechanisms do not exist at sea and this is one reason why the dry deposition velocity of NO_2 on land will be higher than at sea. During winter, deciduous forests have no leaves and consequently the surface resistance will be higher during winter. This is also true for coniferous forests because the transfer velocity through the stomata is a function of temperature. During relatively low temperature and light intensity conditions in winter, the biological activity will be reduced and the surface resistance will therefore be relatively high for these gases. Such effects will not exist at sea.

Particles

The surface resistance for particles is assumed to be negligible both for sea and land surfaces, i.e. that all particles that reach the surface will be taken up by the sea water.

Fluxes

The flux of gases to a surface is calculated from

$$F = -v_e(c_a - c_w{}^*) \tag{16}$$

where F is the flux for reference height z_{ref} (mol m^{-2} s^{-1}). The flux is by definition negative if the compound is removed from the atmosphere. In this equation c_a is the air concentration above the sea surface (mol m^{-3}) and $c_w{}^*$ is the theoretical gas phase concentration, which would be in equilibrium with the concentration of dissolved gas at the sea–air interface (mol m^{-3} air). If c_a is larger than $c_w{}^*$, dry deposition takes place. In the opposite case emission takes place.

The dry deposition velocity is defined by

$$v_d \equiv \frac{F}{c_a} \tag{17}$$

It can be seen from this equation that v_d is only equal to v_e if $c_w{}^*$ is zero. This is the case for most gases, except for NH_3. This holds for sea water and for agricultural crops.

$c_w{}^*$ of NH_3 in sea water can be found from [Asman et al. 1994b]:

$$c_s{}^* = \frac{[NH_x]}{RTH_{NH_3}\left(\dfrac{1}{\gamma_{NH_3}} + \dfrac{10^{-pH}}{\gamma_{NH_4}K_{NH_4}}\right)} \tag{18}$$

where

c_s^* = NH$_3$ concentration in air, which would be in equilibrium with [NH$_x$] (mol m^{-3})

[NH$_x$] = NH$_x$ concentration in the surface (mol m^{-3})

γ_{NH_3} = activity coefficient NH$_3$.H$_2$O

γ_{NH_4} = activity coefficient NH$_4^+$

R = gas constant (8.2075 × 10^{-2} atm l mol^{-1} K^{-1})

H_{NH_3} = Henry's law constant for NH$_3$ (mol l^{-1} atm^{-1})

pH   = pH water in surface, which is a measure of the activity of H$^+$ in the surface solution

K_{NH_4} = dissociation constant for NH$_4^+$ (mol l^{-1}).

The value of c_s^* increases strongly with temperature due to the temperature dependence of H_{NH3} and K_{NH4}.

The activity coefficient of NH$_3$.H$_2$O in sea water is approximated by the following equation, which is valid for an NaCl solution of 25°C [Randall and Failey, 1927]:

$$\gamma_{NH_3} = 1 + 0.085 \, I \tag{19}$$

The ionic strength I for sea water as a function of salinity can be found from [Lyman and Fleming, 1940]

$$I = 0.00147 + 0.01988 \, S + 2.08357 \times 10^{-5} \, S^2 \tag{20}$$

Moreover, the relationship is not valid if sea water is diluted considerably with water for which the relative contribution of the different ions differs markedly from that in sea water (e.g. in estuaries).

The activity coefficient for NH$_4^+$ in sea water at 25°C can be found by using an ion pairing model [Millero and Schreiber, 1982; Asman et al., 1994b]

$$\gamma_{NH_4^*} = 0.883 - 0.0768 \, \ln(sal) \tag{21}$$

Asman et al. [1994b] found such high NH$_3$.H$_2$O concentrations in the North Sea, that dry deposition is reduced substantially. In a few cases an upward flux was noted. Quinn et al. [1988] describe such NH$_3$ emissions in the Pacific Ocean. In general, marine (endogenous) NH$_3$ emission at sea is necessary to explain NH$_4^+$ concentrations in precipitation over the oceans [Dentener and Crutzen, 1994].

Table 2.5 shows exchange velocities for gases and particles. Table 2.5 shows also the rate at which the concentration of the compounds will be removed (% h^{-1}) by dry deposition if no substantial concentration were available in sea water. The dry deposition velocity is also shown for an average land surface (z_{0m} = 0.3 m). Table 2.5 shows that dry deposition velocities are much lower over the sea than over land. This is caused by the difference in surface roughness as well as by the different properties of the surface (on land biological processes in vegetation can be important). For this reason, specific dry deposition velocities should be used for sea areas to calculate the dry deposition to the sea.

TABLE 2.5. Computed exchange velocities and removal rates of nitrogen compounds to sea and land surface at a wind velocity of 5 m s^{-1} at 10 m height and a mixing height of 1000 m [Asman et al., 1994c].

Compound	Sea		Land	
	ν_e (mm s^{-1})	Removal rate (% h^{-1})	ν_e (mm s^{-1})	Removal rate (% h^{-1})
NH$_3$	7.6	2.7	22	7.9
NH$_4^+$ aerosol	0.2	0.07	1.23	0.44
NO	3.5×10^{-4}	1×10^{-4}	0.98	0.35
NO$_2$	2.2×10^{-3}	8×10^{-4}	6.05 (day);	2.2
			2.74(night)	1.0
HNO$_2$	6.9	2.5	8.67	3.1
HNO$_3$	6.4	2.3	65.1	23
NO$_3$ radical	6.6	2.4	65.1	23
N$_2$O$_5$	6.1	2.2	65.1	23
PAN	1.0	0.36	1.94	0.70
NO$_3^-$ aerosol	0.2	0.07	1.23	0.44

The dry deposition velocities in Table 2.5 are presented more accurately than actually known. The uncertainty is at least 50%.

2.8. Model Results and Discussion

Processes that play a role in atmospheric transport models have been presented and discussed. In principle, an atmospheric transport model can now be constructed by putting the processes for each component in a differential equation of type (1). In this way, a set of equations is obtained, which has to be integrated to find the concentration/deposition at a receptor point at sea. There exist many different types of atmospheric transport models. They are different in the sense that they are either Eulerian or Lagrangian, that contrasting numerical methods are used to solve the equations, different sets of chemical reactions are used, different horizontal and vertical resolution is used etc. The choice of the model type depends on many factors: the compounds and areas of interest, the location of important source areas compared to the location of the areas for which the deposition should be calculated, the length of the period of interest etc. A factor that can be very important is the availability of computer resources, because the necessary calculations take much time. A way to avoid excessive cpu time is to calculate the deposition for a limited set of different situations (classes). All situations are then attributed to one of these classes and the average deposition is then found from the values for each class weighted with the frequency of occurrence of that class. This type of model is called a statistical model [see e.g. Asman and Runge, 1991].

Before applying a model, its results must be verified with measurements. An example for such verification for the ACDEP model for the wet deposition of NH$_4^+$, which is shown in Figure 2.9.

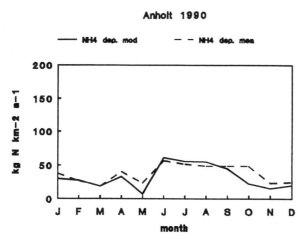

Figure 2.9. Comparison of modeled and measured wet deposition of NH_x for Anholt in 1990.

Table 2.6 shows results of the calculations with the ACDEP model for the Kattegat for the year 1990. The model results were in reasonable agreement with measured concentrations and wet deposition data [Asman et al. 1994a]. The largest part of the atmospheric nitrogen input is attributed to wet deposition. Wet deposition of NH_x contributes 42% to the total nitrogen input, while NO_y contributes 30%. The NH_x contribution to the dry deposition is 17% while the NO_y contribution is 11%. NH_3 and HNO_3 contribute most to the dry deposition, because of their high deposition velocities despite their lower concentrations relative to NO_2, PAN or the nitrogen-containing aerosols. The contribution of NH_3 and HNO_3 to the wet depo-

TABLE 2.6. Deposition of nitrogen compounds to the Kattegat (kg N km^{-2} yr^{-1})[1] [Asman et al., 1994a].

Compound	dry	wet	total
NH_3	167.2	64.2	231.4
NH_4^+ aerosol	16.8	340.3	357.1
NO	1.6	0.0	1.6
NO_2	18.1	0.0	18.1
HNO_2	0.5	0.0	0.5
HNO_3	46.8	90.1	136.9
HNO_4	0.5	1.0	1.5
NO_3 radical	0.4	1.2	1.6
N_2O_5	7.4	15.8	23.2
PAN	3.6	0.0	3.6
Organic NO_3^-	1.4	0.0	1.4
NO_3^- aerosol	6.9	174.9	181.8
Total N	271.2	687.5	958.7

The numbers are given much more accurately than actually known.

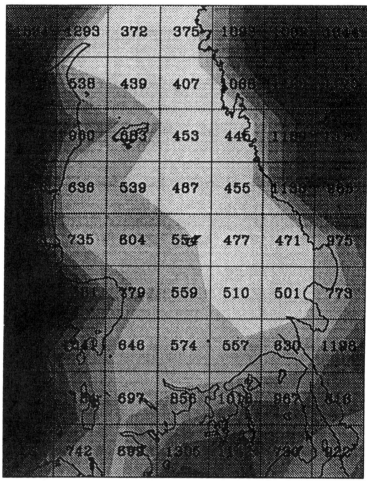

Figure 2.10. Total NH$_x$ deposition in Kattegat (kg N km^{-2} yr^{-1}).

sition is also important, but the contribution of the aerosols dominates. Table 2.6 illustrates again that wet deposition is a more important removal process for aerosols than dry deposition. The geographical distributions of the total deposition (the sum of wet and dry deposition) of NH$_x$ and NO$_y$ are given in Figures 2.10 and 2.11 showing lower depositions at sea than on land. This difference between land and sea results is to a large part caused by the lower amount of precipitation at sea, and also by lower dry deposition velocities at sea, and also because most sources are land-based, which leads to higher concentrations on land. Model results show that Denmark contributes 35% of the total NH$_x$ deposition to the Kattegat, but only 7% to the total NO$_y$ deposition. In fact, Germany is the largest contributor to the total NO$_y$ deposition to the Kattegat with about 35%. Detailed results of the calculations are presented in Asman et al. [1994a].

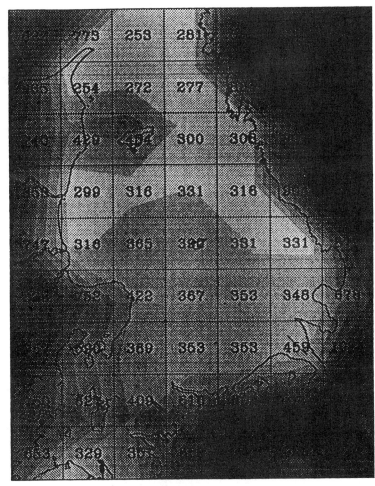

Figure 2.11. Total NO_y deposition in Kattegat (kg N km^{-2} yr^{-1}).

The contribution of the atmospheric input of nitrogen to the Kattegat is about 20–30% of the total input including the transport from other sea areas, runoff etc. (see chapter 11).

In general, wet deposition is more important than dry deposition at sea, whereas near source areas on land dry deposition can be more important for compounds with high dry deposition velocities.

Lindfors et al. [1991] estimated the nitrogen deposition to the Baltic Sea from measured concentrations and measured meteorological data. They found a nitrogen deposition to the South Baltic, Central Baltic and North Baltic of 1140±350, 910±270 and 650±280 kg N km^{-2} yr^{-1}, respectively. Van Jaarsveld [1992] gives an overview of estimates of the nitrogen deposition to the North Sea. He gives as best estimates a total deposition of NH_x (sum of dry and wet deposition) of 270 kg N km^{-2} yr^{-1} and a total deposition of NO_y of 380 kg N km^{-2} yr^{-1}. This gives a total N deposition of 650 kg N km^{-2} yr^{-1}. The wet deposition of NH_x to remote oceanic areas is 40–50 kg N km^{-2} yr^{-1} and the wet deposition of NO_y in these areas is 30–40 kg N km^{-2} yr^{-1} [Galloway, 1985; Buijsman et al., 1991; Duce et al, 1991].

Acknowledgments. This work was partly funded by the Danish Environmental Protection Agency, Copenhagen, Denmark, within the project Danish Marine Research Programme 90. This work is partly based on research carried out by the following persons: Peter Allerup, Jes Jørgensen, Henning Madsen, Søren Overgaard, Flemming Vejen (Danish Meteorological Institute, Copenhagen, Denmark), Sven-Erik Gryning, Anna Maria Sempreviva, Poul Hummelshøj, Niels Otto Jensen (Risø National Laboratory, Roskilde, Denmark), Ole Hertel, Ruwim Berkowicz, Erik H. Runge, Jesper Christensen, Mads Hovmand (National Environmental Research Institute). Trajectories used in the model calculations were provided by the Meteorological Synthesizing Centre-West, European Monitoring and Evaluation Programme at the Norwegian Meteorological Institute, Oslo, Norway.

References

Allen, A. G., R. M. Harrison, and J. W. Erisman, Field measurements of the dissociation of ammonium nitrate and ammonium chloride aerosols, *Atmos. Envir.*, 23, 1591–1599, 1989.

Allerup, P. J. Jørgensen, H. Madsen, S. Overgaard, F. Vejen, and W. A. H. Asman, Precipitation over Kattegat (in Danish), *Havforskning fra Miljøstyrelsen, 14,* 127 pp., Danish Environmental Protection Agency, Copenhagen, 1992.

Arya, S. P., *Introduction to Micrometeorology*, Academic Press, San Diego, 1988.

Asman, W. A. H., Ammonia emission in Europe: Updated emission and emission variations. *Rep. 228471008*, National Institute of Public Health and Environmental Protection (RIVM), Bilthoven, The Netherlands, 1992.

Asman, W. A. H., Parameterization of below-cloud scavenging of highly soluble gases under convective conditions, *Atmos. Envir.*, 29, 1359–1368, 1995.

Asman, W. A. H., and E. H. Runge, Atmospheric deposition of nitrogen components to Denmark and surrounding sea areas – A preliminary estimate, *Rep. NFAC R-91-1*, National Environmental Research Institute, Roskilde, Denmark, 1991.

Asman, W. A. H., and J. A. van Jaarsveld, A variable-resolution transport model applied for NH_x for Europe, *Atmos. Envir.*, 26A, 445–464, 1992.

Asman, W. A. H., B. Drukker, and A. J. Janssen, Modelled historical concentrations and depositions of ammonia and ammonium in Europe, *Atmos. Envir.*, 22, 725–735 1988.

Asman, W. A. H., E. H. Runge, and N. A. Kilde, Emission of NH_3, NO_x, SO_2 and NMVOC to the atmosphere in Denmark (in Danish), *Havforskning fra Miljøstyrelsen, 19,* 105 pp., Danish Environmental Protection Agency, Copenhagen, 1993.

Asman, W. A. H., R. Berkowicz, J. Christensen, O. Hertel, and E. Runge, Atmospheric deposition of nitrogen components to the Kattegat (in Danish), *Havforskning fra Miljøstyrelsen, 37,* 115 pp., Danish Environmental Protection Agency, Copenhagen, 1994a.

Asman, W. A. H., R. M. Harrison, and C. J. Ottley, Estimation of the net air-sea flux of ammonia over the southern bight of the North Sea, *Atmos. Envir.*, 28, 3647–3654, 1994.

Asman, W. A. H., L. L. Sørensen, R. Berkowicz, K. Granby, H. Nielsen, B. Jensen, and E. Runge, C. Lykkelund, S. E. Gryning, and A. M. Sempreviva, Dry deposition processes (in Danish), *Havforskning fra Miljøstyrelsen*, 35, 199 pp., Danish Environmental Protection Agency, Copenhagen, 1994c.

Baldwin, A. C., and D. M. Golden, Heterogeneous atmospheric reactions: Sulphuric acid as a tropospheric sink, *Science* 206, 562–563, 1979.

Bates, R. G., and G. D. Pinching, Dissociation constant of aqueous ammonia at 0 and 50°C from E.m.f. studies on the ammonium salt of a weak acid., *Am. Chem. J.*, 72, 1393–1396, 1950.

Brutsaert, W., *Evaporation into the atmosphere*, Kluwer, Dordrecht, The Netherlands, 1991.

Buijsman, E., J. F. M. Maas, and W. A. H. Asman, Anthropogenic NH_3 emissions in Europe, *Atmos. Envir.*, 21, 1009–1022, 1987.

Buijsman, E., P. J. Jonker, W. A. H. Asman, and T. B. Ridder, Chemical composition of precipitation collected on a weathership on the North Atlantic, *Atmos. Envir.*, 25A, 873–883, 1991.

Dasgupta, P. K., and S. Dong, Solubility of ammonia in liquid water and generation of trace levels of standard gaseous ammonia, *Atmos. Envir.*, 20, 565–570, 1986.

Davidson, C. I., and Y.-L. Wu, Dry deposition of particles and vapors, in *Acidic Precipitation*, Vol. 3. *Sources, Deposition, and Canopy Interactions*, edited by S. E. Lindberg, A. L. Page, and S. A. Norton, pp. 103–215, Springer, New York, 1990.

Dentener, F. J., and P. J. Crutzen, A three dimensional model of the global ammonia cycle. *J. Atmos. Chem.*, 19, 331–369, 1994.

Duce, R. A., P. S. Liss, J. T. Merrill, E. L. Atlas, P. Buat-Ménard, B. B. Hicks, J. M. Miller, J. M. Prospero, R. Arimoto, T. M. Church, W. Ellis, J. N. Galloway, L. Hansen, T. D. Jickells, A. H. Knap, K. H. Reinhardt, B. Schneider, A. Soudine, J. J. Tokos, S. Tsunogai, R. Wollast, and M. Zhou, The atmospheric input of trace species to the world ocean, *Global Biogeochem. Cycles*, 5, 193–259, 1991.

Dugdale, R. A., Nutrient limitations in the seas: dynamics, identification and significance, *Limnol. Oceanogr.*, 685–695, 1967.

Erisman, J. W., A. W. M. Vermetten, W. A. H. Asman, A. Waijers-Ypelaan, and J. Slanina, Vertical distribution of gases and aerosols: The behaviour of ammonia and related components in the lower atmosphere, *Atmos. Envir.*, 22, 1153–1160, 1988.

Fitzgerald, J. W., Approximation formulas for the equilibrium size of an aerosol particle as a function of its dry size and composition and the ambient relative humidity, *J. Appl. Met.*, 14, 1044–1049, 1975.

Galloway, J. N., The deposition of sulfur and nitrogen from the remote troposphere, in *The Biogeochemical Cycling of Sulfur and Nitrogen in the Remote Atmosphere*, edited by J. N. Galloway, R. J. Charlson, M. O. Andreae, and H. Rodhe, pp. 143–175, 1985.

Garratt, J. R., The internal boundary layer – a review, *Boundary Layer Met.*, 50, 171–203, 1990.

Gery, M. W., G. Z. Whitten, J. P. Killus, and M. C. Dodge, A photochemical kinetics mechanism for urban and regional computer modelling, *J. Geophys. Res.*, 94D, 12925–12956, 1989.

Gryning, S. E., Wind, turbulence and boundary layer height over Kattegat (in Danish), *Havforskning fra Miljøstyrelsen*, 21, 57 pp., Danish Environmental Protection Agency, Copenhagen, 1993.

Harrison, R. M., and A.-M. N. Kitto, Estimation of the rate constant for the reaction of acid sulphate aerosol with NH_3 gas from the atmospheric measurements. *J. Atmos. Chem.*, 15, 133–143, 1992.

Hertel, O., R. Berkowicz, W. A. H. Asman, J. Christensen, and L. L. Sørensen, Description of chemical processes in the atmosphere (in Danish), *Havforskning fra Miljøstyrelsen*, 24, 45 pp., Danish Environmental Protection Agency, Copenhagen, 1993.

Hinds, W. C., *Aerosol Technology*, John Wiley & Sons, New York, 1982.

Huntzicker, J. J., R. A. Cary, and C.-S. Ling, Neutralization of sulphuric acid aerosol by ammonia, *Envir. Sci. Technol., 14,* 819–824, 1980.

Kames, J., S. Schweighoefer, and U. Schurath, Henry's law constant and hydrolysis of peroxyacetyl nitrate (PAN), *J. Atmos. Chem., 12,* 169–180, 1991.

Krell, U., and E. Roeckner, Model simulation of the atmospheric input of lead and cadmium into the North Sea, *Atmos. Envir., 22,* 375–381, 1988.

Larsen, S. E., P. Hummelshøj, N. O. Jensen, J. B. Edson, G. de Leeuw, and P. G. Mestayer, Deposition of airborne particles to the sea surface (in Danish), *Havforskning fra Miljøstyrelsen, 47,* 79 pp., Danish Environmental Protection Agency, Copenhagen, 1994.

Larsen, S. E., P. Hummelshøj, N. O. Jensen, J. B. Edson, G. de Leeuw, and P. G. Mestayer, Dry deposition of particles to water surfaces. *Ophelia, 42,* 193–204, 1995.

Levine, S. Z., and S. E. Schwartz, In-cloud and below-cloud scavenging of nitric acid vapor, *Atmos. Envir., 16,* 1725–1734, 1982.

Lindfors, V., S. M. Joffre, and J. Damski, Determination of the wet and dry deposition of sulphur and nitrogen compounds over the Baltic Sea using actual meteorological data, *FMI Contr. No. 4,* Finnish Meteorological Institute, Helsinki, Finland, 1991.

Liss, P. S., Gas transfer: experiments and geochemical implications, in *Air-Sea Exchange of Gases and Particles,* edited by P. S. Liss, and W. G. N. Slinn, pp. 241–298, Reidel, Dordrecht, The Netherlands, 1983.

Lyman, J. and R. H. Fleming, Composition of seawater, *J. Mar. Res. 3,* 134–146, 1940.

McMurry, P. H., H. Takano, and G. R. Anderson, Study of the ammonia (gas)-sulphuric acid (aerosol) reaction rate, *Envir. Sci. Technol., 17,* 347–352, 1983.

Millero, F. J., and D. R. Schreiber, Use of the ion pairing model to estimate activity coefficients of natural waters, *Am. J. Sci., 282,* 1508–1540, 1982.

Möller, U., and G. Schumann, Mechanisms of transport from the atmosphere to the earth's surface, *J. Geophys. Res., 75,* 3013–3019, 1970.

Mozurkewich, M., The dissociation constant of ammonium nitrate and its dependence on temperature, relative humidity and particle size, *Atmos. Envir., 27A,* 261–270, 1993.

Ottley, C. J., and R. M. Harrison, The spatial distribution and particle size of some inorganic nitrogen, sulphur and chlorine species over the North Sea, *Atmos. Envir., 26A,* 1689–1699, 1992.

Pacyna, J. M., S. Larssen, and A. Semb, European survey for NO_x emissions with emphasis on Eastern Europe, *Atmos. Envir., 25A,* 425–439, 1991.

Paerl, H. W., and M. L. Fogel, Isotopic characterization of atmospheric nitrogen inputs as sources of enhanced primary production in coastal Atlantic Ocean waters, *Mar. Biol.,* in press, 1994.

Peters, L. K., Gases and their precipitation scavenging in the marine atmosphere, in *Air-Sea Exchange of Gases and Particles,* edited by P. S. Liss, and W. G. N. Slinn, pp. 173–240, Reidel, Dordrecht, The Netherlands, 1983.

Petersen, G., H. Weber, and H. Grassl, Modelling the transport of trace metals from Europe to the North Sea and the Baltic Sea, in *Air Pollution Modeling and its Application,* edited by J. M. Pacyna, and B. Ottar, pp. 581–583, NATO ASI Series, Kluwer, Dordrecht, The Netherlands, 1988.

Pio, C. A., and R. M. Harrison, The equilibrium of ammonium chloride aerosol with gaseous hydrochloric acid and ammonia under tropospheric conditions, *Atmos. Envir., 21,* 1243–1246, 1987.

Quinn, P. K., R. J. Charlson, and T. S. Bates, Simultaneous observations of ammonia in the atmosphere and ocean, *Nature, 335,* 336–338, 1988a.

Quinn, P. K., R. J. Charlson, and W. H. Zoller, Ammonia, the dominant base in the remote marine troposphere: A review, *Tellus, 39B,* 413–425, 1988b.

Quinn, P. K., T. S. Bates, J. E. Johnson, D.S. Covert, and R. J. Charlson, Interactions between the sulfur and reduced nitrogen cycles over the Central Pacific Ocean, *J. Geophys. Res.*, 95, 16405–16416, 1990.

Randall, M., and C. F. Failey, The activity coefficient of gases in aqueous salt solutions, *Chem. Rev.*, 4, 271–284, 1927.

Rendell, A. R., C. J. Ottley, T. D. Jickells, and R. M. Harrison, The atmospheric input of nitrogen species to the North Sea, *Tellus, 45B*, 53–63, 1993.

Robbins, R. C., and R. D. Cadle, Kinetics of the reaction between gaseous ammonia and sulphuric acid droplets in an aerosol, *Phys. Chem.* 62, 469–471, 1958.

Ryther, J. H., and W. M. Dunstan, Nitrogen, phosphorus and eutrophication in the coastal marine environment, *Science 171*, 1008–1112, 1971.

Schwartz, S. E., and W. H. White, Solubility equilibria of the nitrogen oxides and oxyacids in dilute solution, in *Advances in Environmental Science and Engineering*, Vol. 4, edited by J. R. Pfafflin, and E. N. Ziegler, pp. 1–45, Gordon/Breach, New York, 1981.

Sehmel, G. A., and S. L. Sutter, Particle deposition rates on a water surface as a function of particle diameter and air velocity, *J. Rech. Atmos.* 8, 911–920, 1974.

Sievering, H., Profile measurements of particle mass transfer at the air-water interface, *Atmos. Envir.*, 15, 123–129, 1981.

Sievering, H., Small-particle dry deposition on natural waters: Modeling uncertainty, *J. Geophys. Res.*, 89, 9679–9681, 1984.

Smith, M. H., P. M. Park, and I. E. Consterdine, North Atlantic aerosol remote concentrations measured at a Hebridean coastal site, *Atmos. Envir.*, 25A, 547–555, 1991.

Slinn, W. G. N., Air-to-sea transfer of particles, in *Air-Sea Exchange of Gases and Particles*, edited by P. S. Liss, and W. G. N. Slinn, pp. 299–405, Reidel, Dordrecht, The Netherlands, 1983.

Slinn, S. A., and W. G. N. Slinn, Predictions for particle deposition on natural waters. *Atmos. Envir.*, 14, 1013–1016, 1980.

Stelson, A. W., S. K. Friedlander, and J. H. Seinfeld, A note on the equilibrium relationship between ammonia and nitric acid and particulate ammonium nitrate, *Atmos. Envir.* 13, 369–371, 1979.

Stelson, A. W., and J. H. Seinfeld, Relative humidity and temperature dependence of the ammonium nitrate dissociation constant, *Atmos. Envir.*, 16, 903–922, 1982a.

Stelson, A. W., and J. H. Seinfeld, Relative humidity and pH dependence of the vapor pressure of ammonium nitrate-nitric acid solutions at 25°C, *Atmos. Envir.*, 16, 993–1000, 1982b.

Stelson, A. W., and J. H. Seinfeld, Thermodynamic prediction of the water activity, NH_4NO_3-$(NH_4)_2SO_4$-H_2O system at 25°C, *Atmos. Envir.*, 16, 2507–2514, 1982c.

Stull, R. B., *An Introduction to Boundary Layer Meteorology*, Kluwer, Dordrecht, The Netherlands, 1988.

van Jaarsveld, J. A., Estimating atmospheric inputs of trace constituents to the North Sea: Methods and results, in *Air Pollution Modeling and its Application IX*, edited by H. van Dop, and G. Kallos, pp. 249–258, Plenum, New York, 1992.

Williams R. M., A model for the dry deposition of particles to natural water surfaces, *Atmos. Envir.*, 16, 1933–1938, 1982.

3

Water Masses, Stratification and Circulation

Jacob Steen Møller

3.1. Introduction

The aim of this chapter is to describe the physical oceanographic processes that may play a role in eutrophication. The manifestation of adverse effects of eutrophication in coastal marine environments is closely related to hydrographic processes. The term hydrographic processes is used here to mean physical processes such as advection, the development of stratification, establishment of fronts and the mixing of water masses. The salinity and temperature stratification, which is present in many coastal waters, results in a spatial separation of photosynthetic and mineralization processes which can lead to oxygen depletion of the lower layers of the water column [Fenchel, 1992].

It is important to note that the oxygen depletion is not caused by the eutrophication itself, but that oxygen depletion requires hydrographic conditions to develop that favor stagnant or isolated water masses. Such a spatial separation of primary production and mineralization is not unique but is found in many marine areas around the world. Among the more extreme examples of areas exhibiting this phenomenon are the Black Sea and the North Sea.

In the Black Sea, the hydrographic conditions have caused oxygen depletion independently of the recent increase in nutrient loads. The Black Sea receives a large riverine inflow, where the surplus of water leaves the Black Sea through the Bosporus Strait (Figure 3.1). In the Bosporus, a pronounced two-layer flow system is maintained by the surplus of water from the Black Sea and the inflow of saline Mediterranean water entering through first the Dardanelles and then the Marmara Sea. The flow through the Bosporus is due to a balance between the surface water slope (the barotropic gradient) which forces the surface water out of the Black Sea

Eutrophication in Coastal Marine Ecosystems
Coastal and Estuarine Studies, Volume 52, Pages 51–66
Copyright 1996 by the American Geophysical Union

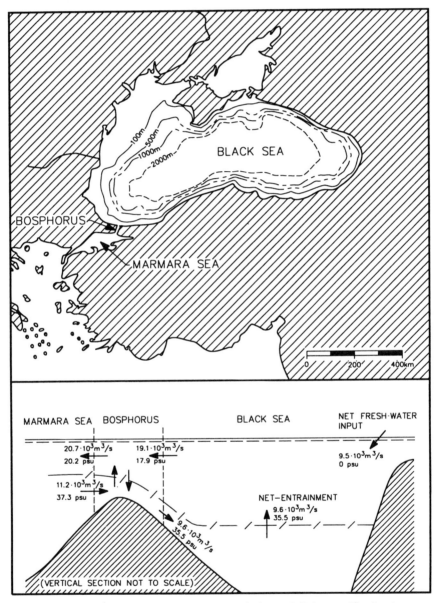

Figure 3.1. The Black Sea. Water exchange of the Black Sea. Salinity stratification is responsible for natural oxygen depletion of the strata below the halocline. Data elaborated from
Ünlüata [1990].

and the pressure gradient at the lower layer caused by the density difference (the
baroclinic gradient) between the brackish Black Sea water and the saline Mediterranean water in the Marmara. The stratification of the Black Sea is thereby main-

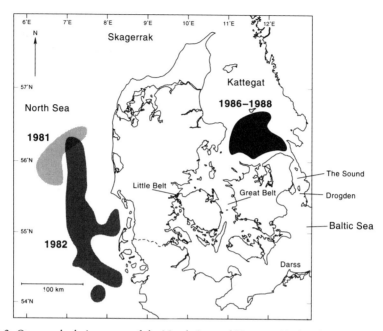

Figure 3.2. Oxygen depletion areas of the North Sea and Kattegat [Richardson, 1989].

tained by the balance of the saline water inflow through the Bosporus and the vertical mixing of the saline bottom water into the brackish surface water. The vertical mixing is primarily caused by wind stirring. Phytoplankton and detritus from the upper water layer of the Black Sea settle to the bottom layer where they are degraded and consume oxygen; in addition the sediment will also sustain a potential oxygen demand due to degradation of settled material. The supply of oxygen to the lower layer is limited to the advective transport through the Bosporus because the stratification hampers downward mixing (entrainment) of oxygen-rich surface water into the lower layer. The consumption of oxygen thus exceeds the oxygen supply and oxygen depletion becomes a permanent situation below the pycnocline.

The North Sea has experienced considerable eutrophication during the last decades and, in recent years, oxygen depletion has been reported [e.g., Richardson, 1989]. The oxygen depletion has been observed in the parts of the North Sea where temperature stratification develops during the summer (Figure 3.2). The solar heating of the surface water stabilizes the water column and creates a stratification which hampers the downward transport of oxygen and thus enhances an oxygen depletion. The extension of the depleted area is, among other factors, determined by the extension of the stratification, maintained through a balance between the stabilizing forces (heat input) and the destabilizing forces (tidal and wind-mixing) [e.g., Pedersen, 1994].

The North Sea and the Black Sea illustrate the importance of the hydrographic conditions for the development of eutrophication effects. Also the eutrophication processes of the Danish Straits and coastal waters are strongly influenced by the hydrography. In the following, the basic physical phenomena of the Danish Straits are described, and it is seen that many of the hydrographic processes identified in the Black Sea and North Sea also prevail in these coastal areas.

3.2. Advection and Mixing

Transport and mixing of water and matter in the sea are described by the general equations of continuity and motion. In their exact form, these equations contain terms that express the transport of momentum, heat and mass. Generally, however, it is not possible to solve the equations and express the transport and mixing of matter in the ocean without introducing simplifying assumptions. One of the reasons for this is the presence of nonlinear terms in the equations. The equations can only be solved by direct calculation or by integration using a computer code when the nonlinear terms can be either neglected or determined by introducing suitable approximations [Rodi 1980 and 1994].

It is the art of physical oceanographic science to assess and choose suitable simplifications that allow a rational analysis of hydrographic processes without "spoiling" the result. The choice of approximations is dependent upon the problem to be analyzed. In particular, it is important to analyze the temporal and spatial scales of the (biological) system. As an illustration, we can first look at the resuspension of particulate matter in a coastal environment and then at the renewal of bottom water of the deep basins of the Black Sea. It is apparent that the spatial and temporal scales of the two processes are significantly different. In the first case (resuspension) the detailed turbulence structure close to the sea bottom must be described at a time scale taking the wave period into consideration, while in the second case (renewal) such a detailed turbulence description is of minor importance. The problem dictates the proper simplifications. A careful consideration of these aspects should, therefore, precede any biological/oceanographical investigation. The following review of the hydrography of the Danish Straits in relation to eutrophication will in more detail illustrate this link between the assessment of the biological problem and the hydrographic analysis.

The Continuity Equation

The change of a property in the sea, (e.g., the concentration of oxygen or the amount of plankton within the photic layer) can be described as the sum of the changes due to advective transport, turbulent transport (dispersion or mixing) and the exchange with other compartments of matter through biological or chemical reactions:

$$\frac{\partial c}{\partial t} = -u_i \frac{\partial c}{\partial x_i} - \frac{\partial u_i' c'}{\partial x_i} + T(c,...) \tag{1}$$

where t is time, x is the Cartesian coordinate, u is the mean (nonturbulent) flow velocity, c is the concentration of matter, u' and c' are the turbulent part of u and c, and T is a term expressing the interaction between compartments. In Eq. 1, the left term expresses the rate of the change of concentration, the first term on the right the advection (transport by nonturbulent motion) of matter with the current, the second term on the right the turbulent transport (dispersion or mixing) of matter and the last term the chemical/biological exchange of the matter.

Depending on the problem, some of the above terms may be neglected in order to simplify the solution. Typically, the biologist will focus on the last term (T), e.g. by investigating the change in growth rate of some algae as a function of the light penetration (the transport terms are neglected). This is perfectly satisfactory when the investigation is carried out in a closed environment (i.e., an aquarium or a test tube) without spatial gradients in concentrations. However, when the investigation is carried out in the field, the transport terms may be dominant for the observed changes. On the other side, the physicists may tend to neglect the last term (T) and focus on the transport processes which, as described above, are well described by the general equations of motion and continuity. The argument shows that it is important to analyze, which of the terms dominate the problem to be addressed, before the investigation is started.

The Density of Sea Water

The density of the sea water is of great importance for the transport processes in the sea. The vertical mixing is reduced when a vertical density gradient (often referred to as interface) is present and the horizontal movement of water is influenced by horizontal density gradients (often referred to as fronts). The density of sea water depends on the salinity and temperature. For depths greater than several hundred meters, the pressure also influences the density. The relationship between the salinity and temperature on the one side and the density on the other has been determined by measurements and is given in the form of an algorithm (Figure 3.3; UNESCO 1981a,b).

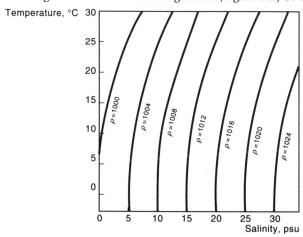

Figure 3.3. The density of sea water, ρ (kg m^{-3}), is a function of temperature and salinity. The Salinity unit psu is defined in UNESCO [1981b], for practical calculations psu is equal to ppt or ‰.

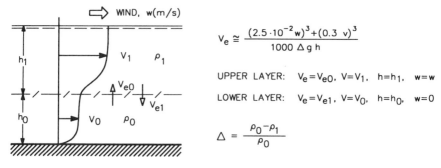

Figure 3.4. A simplified two-layer system and an evaluation of the entrainment velocities based on Pedersen's theory [1986].

Stratification

When the density of the sea water varies with depth, the water column is said to be stratified. The stratification may be caused by heating of the surface water or reduced by salinity due to riverine influence. Other mechanisms such as wind stirring and the action of shear stresses due to bottom friction result in irregular motions (turbulence) of the water. The turbulent motion will result in mixing of water thus working against the forces that cause stratification. Stratification of water masses reduces the mixing between different layers of the water column due to damping of turbulent motions in a density gradient. The turbulent motion varies with current and wind conditions, which may both show great variability. For homogeneous flow, the turbulent kinetic energy eventually is dissipated into heat. In a stratified flow, part of the turbulent kinetic energy is used to mix the layers of different density.

If we consider a situation where the water column is strongly stratified and may be considered as a two-layer system divided by a sharp interface (Figure 3.4), any deformation of the interface will produce a restoring force due to the buoyancy: if bottom water is lifted into the surface layer, gravity will seek to tow it back into the bottom water, and if surface water is pushed into the bottom water, buoyancy will seek to lift the water back into the surface layer. To overcome this mechanism, energy has to be supplied. This energy is transferred from the turbulent kinetic energy of the flow.

Following Pedersen [1986], the vertical mixing process can be described as a two-way entrainment process. When a layer of no movement is overlaid by a turbulent layer, the quiet layer is entrained into the turbulent layer. This process increases the potential energy of the water column by lifting dense water into the surface layer. Similarly, a turbulent bottom layer entrains water from a quiet surface layer. When both layers are turbulent, the process is two-way. The efficiency of the mixing can be assessed by relating the gain in potential energy, due to the mixing, with the production of turbulent kinetic energy of the flow. To illustrate the magnitude of the entrainment we consider a simple two-layer system (Figure 3.4). The mixing between the layers is determined by the turbulence production due to the wind, the velocity shear and the bottom friction. W is the wind speed, h the water layer depth, ν the average velocity of the current, ρ the mean density and ν_e the entrainment velocity.

The entrainment assessment given in Figure 3.4 is only a rough approximation to the problem but it illustrates the very strong (cubic) dependence of the entrainment to the wind speed and the current speed. This strong nonlinear relationship illustrates how short storm events have a large impact on the mixing.

Flow Condition

The advective transport of matter in the sea is governed by the flow conditions. The velocity field enters the continuity equation as shown in Eq.1. The velocity field is determined by the balance of forces acting on the water. This balance is quantified through the momentum equation. The momentum equation expresses the balance between the inertia forces, the pressure and gravity forces, the rotational forces (Coriolis force) and the frictional forces. As it was the case for the continuity equation also the momentum equation cannot be solved in the general case, simplifying assumptions must be introduced depending on the problem to be investigated. Some of the terms may be neglected in order to simplify the solution. In oceanographic problems the friction term may often be neglected, whereas in most coastal problems this term is dominating the problem. It is therefore crucial to address and analyze the dominant terms of the problem before the investigation is started.

In many shallow-water cases the flow is governed by a balance between the bottom friction and the pressure gradient. When the pressure gradient is due to a water level gradient, the flow is said to be barotropic; when the pressure gradient is due to a density gradient, the flow is said to be baroclinic. An example of a barotropically dominated flow is the tidal current in the North Sea. An example of baroclinically dominated flow is the exchange flow between the Marmara Sea and the Black Sea. In the general case, both modes of flow forcing must be considered when the water is stratified.

In many oceanographic cases the rotation of the Earth influences the flow pattern. An example of the influence of the rotation of the Earth is given below in the section on the Skagerrak front. When the rotation of the Earth is entering the problem, the term geostrophic flow is often used expressing that the Coriolis forcing becomes important.

3.3. Hydrography of the Danish Straits

The Danish Straits show examples of the majority of the physical features of relevance when considering eutrophication. The Danish Straits are part of the Baltic Sea and, hydrographically, they form the transition area between the brackish Baltic

Proper and the saline North Sea. In the following these waters are examined and local examples used to illustrate the interaction between hydrography and eutrophication. This discussion also gives the hydrographic basis for the following chapters which deal with biological aspects of the eutrophication processes. An overview of the area is shown in Figure 3.2 where geographical names used in the text can be found.

The conditions inside the strait system are determined by the boundary conditions and the laws of nature [e.g., Møller and Hansen, 1994]. The imposed boundary conditions such as bathymetry, the meteorological forcing, river inflow, tide and surges act through the laws of nature and the resulting hydrographic conditions are considered the output of the Danish Strait system. The hydrographic conditions are described in terms of the salinity and temperature stratification as well as the advection and turbulent mixing of matter.

Bathymetry

The bathymetry of the system is depicted in Figure 5 in chapter 1. Three dominating features should be noted. The hydrographic boundary toward the Baltic Sea is given by the Darss and Drogden sills. The sill depth at Drogden is approx. 8 m and at Darss approx. 17 m. The connection between the North Sea and the Baltic Sea is formed by three channels: the Great Belt, the Sound and the Little Belt. The water depth of the Danish Straits is shallow with typical depths between 10 and 20 m. The boundary between the Kattegat and the Skagerrak in the North Sea is characterized by a sloping bottom toward the deep Norwegian Trench. In the Skagerrak, the depth increases to several hundred meters.

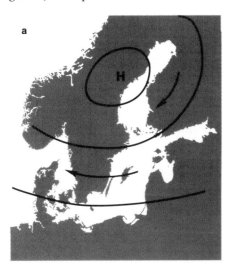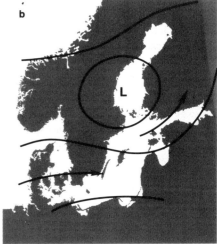

Figure 3.5. Idealized depiction of a) high pressure situation creating a Baltic outflow, b) low pressure situation creating an inflow [Weidemann, 1950].

Tide and Meteorology

The Danish Straits form the transition between the tidal North Sea and the non-tidal Baltic Sea. The tidal influence thus decreases from the Skagerrak toward the Baltic. The flow through the Danish Straits is caused by the water level difference between Kattegat and the western Baltic Sea. Due to the shifting winds, the water level difference is highly variable causing an oscillatory flow through the straits. Even though the net current is northbound (due to the fresh-water surplus of 15,000 m^3 s^{-1} from the Baltic, the instantaneous current fluctuates from northbound to southbound with peak discharges through the channels being 10–15 times the mean discharge. Predictive models for current and water levels, therefore, must include the combined effect of tidal forcing and the meteorological forcing [e.g., Vested et al., 1991]. The meteorological forcing of the system is given by the solar heating and the wind forcing. The wind and atmospheric forcing are the dominating current-generating mechanisms (Figure 3.5) and the wind-induced currents are also the main contributors to the mixing energy in the lower water layer.

Stratification

The waters in the Danish Straits are stratified due to the outflow of brackish water from the Baltic Sea. When the flow is northbound, brackish water is forced through the Straits as a surface current (Figure 3.6). Since the brackish water is of less den-

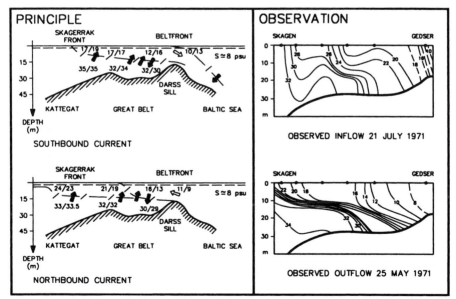

Figure 3.6. Schematic diagram of the current and stratification conditions of the Danish Straits. Modified from *The Belt Project*, Ministry of the Environment [1976].

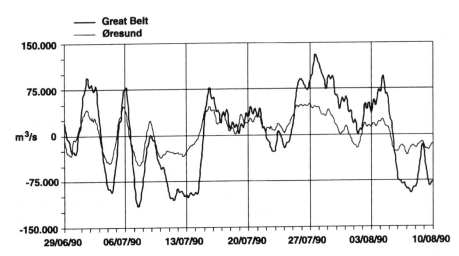

Figure 3.7. Measured discharge through the Great Belt, thick line (at the Great Belt Bridge) and the Sound, thin line (at Drogden) during a six-week period in summer 1990. Modified from Møller and Pedersen [1993].

sity than the water from the Skagerrak, it forms a surface layer. The surface layer depth is basically determined by the sill depth at Drogden and Darss. During southward current the surface water is forced back while undergoing continued mixing with the bottom water. When the more saline water passes the sills it plunges down the slopes.

The Baltic Sea Boundary

The inflow of saline water to the Baltic Sea is a highly intermittent phenomenon (Figure 3.7). The typical duration of an inflow situation is two to four days. However, with irregular intervals, the duration is much longer. In the area where the Kattegat and Belt surface waters meet the Baltic surface water and plunge down the slope, a front (the Belt front) is formed. At this front, the surface salinity rapidly decreases from the Belt surface water salinity (15–25 psu) to the Baltic surface water salinity (8 psu). The discharge under "normal" conditions is, according to Jacobsen [1980], divided between the three channels: the Great Belt, the Sound and the Little Belt in the ratio 7:3:1. If an inflow event continues for a long time, the entire Belt Sea and the Kattegat are flushed with saline North Sea and Skagerrak water and water of high salinity plunges into the Baltic Sea. Under these conditions the total inflow is divided between the Great Belt and the Sound as 8:3, while for the high-salinity water it is 6:5 [Jakobsen, 1995]. Such rare large inflows of saline water play a crucial role for the renewal of the bottom water of the deep basins in the Baltic Sea, whereas the typical and frequent inflow events of a few days' duration renew only the intermediate water layers of the Baltic Sea. A review of the water balance for the Baltic Sea is given by Møller and Hansen [1994], where further references can be found.

The North Sea Boundary

The boundary processes toward the North Sea (Skagerrak) are more complicated than those at the southern boundary. This is due to the more open bathymetry that allows for an internally controlled flow as opposed to the flow over the Drogden and Darss sills where the bathymetry and bottom friction together govern the flow.

In the Skagerrak, the Baltic outflow forms a surface layer of brackish water which, in general, follows the eastern and northern coastline due to the Coriolis force. The interface at the surface between the Baltic outflow and the Skagerrak surface water is called the Skagerrak front. The location of this density front is dependent on the combined influence of tidal and wind forcing. The hydrography of the front shows an annual variation corresponding to the variation in the Baltic outflow and the local wind forcing, [Jakobsen et al. 1994 and Andersson and Rydberg, 1993].

The residual tidal current at the Skagerrak coast of Jutland is eastbound and this, combined with the predominant westerly winds, creates a counter-clockwise (cyclonic) current in the Skagerrak. The saline North Sea water dives under the Baltic outflow when it meets the Skagerrak front. In the central Skagerrak, a doming of Atlantic water and water from the northern and central North Sea is observed. In the dome, water of salinity 35 is raised to levels –20 m or less, [e.g., Poulsen, 1991]. The North Sea water and the Atlantic water mix and enter the Kattegat below the Baltic outflow water. Due to riverine water inflow to the southern North Sea the water with origin here tends to interleave between the Kattegat surface water and the deep inflow from the Skagerrak. Water from the southern North Sea holds higher concentrations of nutrients due to the river runoff; this has been identified as interleaving water in the Kattegat with higher nutrient contents than the bottom water. The water from the southern North Sea can be distinguished from the Skagerrak and other North Sea water by its content of Gelbstoff [Jakobsen et al. 1994].

Mixing and Effects of Stratification

Due to mixing between the layers, the surface salinity increases from the south toward the north. An observer following a northbound surface current through the Danish Straits will thus experience increasing salinity over time due to this mixing. Because the salinity increases toward the north, measurements taken at a fixed point in the Belts will show increasing salinity over time during southward flow and decreasing salinity over time when the flow is northward. In general the energy transfer from the wind to the sea is proportional to the cube of the wind speed, whereby the mixing between the layers is dependent on the third power of the wind speed, Figure 3.4. The wind speed cubed shows a more extreme variation than the wind speed itself.

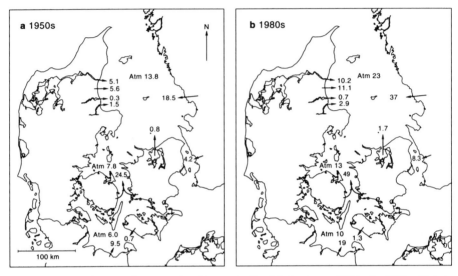

Figure 3.8. Distribution of nitrogen loading to the Danish Straits. a: 1950s, b: 1980s. Unit is 1000 tonnes of nitrogen per year. Atm. is the atmospheric source. From Hansen et al. [1994].

General Circulation

The general water circulation in a two-layer model of the Danish Straits and the nutrient transports and sources to the system are described by Hansen et al. [1990 and 1994]. Based on a review of available salinity and current observations, they determined the mean discharges and vertical exchanges between the different parts of the Danish Straits. The findings are presented in chapter 11.

These data were used as input to a model for the nitrogen and oxygen balance of the bottom water. In Figure 3.8, the increase in the estimated nitrogen loading to the system from the 1950s to the 1980s is illustrated. It is noted that the contributions from the Skagerrak and Baltic Sea are not stated in Figure 3.8. This is due to the complicated stratification and current patterns which make the interpretation of the loading figure difficult. Hansen et al. [1994] conclude that the nitrogen concentration at the boundary toward the Skagerrak and hence the contribution to the nutrient balance of Kattegat have not changed significantly between the 1950s and the 1980s. The contribution from the Skagerrak is approximately 160,000 tons of inorganic nitrogen per year. Hansen et al. [1994] argue that the contribution from the Baltic Sea has increased by a factor 1.7 between the 1950s and the 1980s. The contribution from the Baltic Sea during the 1980s is approximately 20,000 tons of inorganic nitrogen per year. Data on phosphorus loading are insufficient to estimate changes over time. In general, it is advised to investigate both the nitrogen and the phosphorus loading as part of eutrophication studies. Model results suggest that the significant increase in nitrogen loading relates directly to the observed eutrophication effect in the Danish Straits.

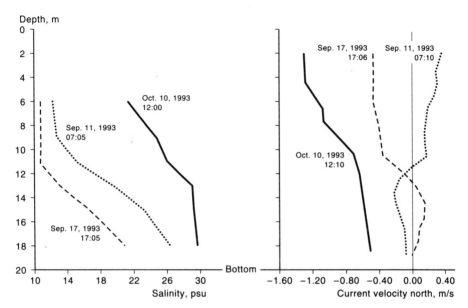

Figure 3.9. Example of salinity and current profiles at the Great Belt Bridge on three days during fall 1993. Note the influence of stratification on the velocity profile. Unpublished data from A/S Storebæltsforbindelsen.

Current and the Effects of Stratification

Because of the variable balance between the barotropic and the baroclinic component, the current can be different in speed and direction for various depths. The mean current in the upper layer is directed out of the Baltic, whereas the mean current in the bottom layer is directed toward the Baltic. This is due to the general estuarine circulation. It should, however, be noted that the instantaneous current profile may be entirely different due to the combined action of wind, density gradients and tidal forcing (see Figure 3.9).

On October 10, 1993, the current profile is barotropically dominated. The flow is toward the Baltic (southward) at all depths. On September 17, 1993, the current is toward the Baltic (southward) for the surface layer and toward Kattegat (northward) for the bottom layer. The barotropic pressure gradient, caused by the water level being higher in Kattegat than in the Baltic, forces the surface water toward the south, whereas the baroclinic pressure gradient, which is caused by a tilting interface between the two layers, balances the barotropic gradient in the lower layer and forces the bottom water to flow toward north. On September 11, 1993, the situation is the opposite. The water level is higher in the Baltic than in the Kattegat, forcing the surface water to flow toward north, while the baroclinic gradient due to the salinity being larger in the Kattegat than further south, together with a tilting interface, forces a southward bottom current.

Scaling considerations

The large variability in current conditions, stratification and mixing conditions both in time and in space, has great influence on the interpretation of the eutrophication mechanisms. The Kattegat may be considered a two-layer system *on an average*, but when looked upon at a time scale of days and hours, the system is more complicated. It is, therefore, important to adjust the time scale of the hydrographic investigation to the time scale of the biological problem in question. If, for example, the long-term trends in primary production in the southern Kattegat are investigated, it may be appropriate to pool all data from the southern Kattegat for each year and analyze it in combination with the averaged hydrographic conditions. If, on the other hand, the spring phytoplankton bloom, which develops over a few days, is examined, the hydrographic processes must be resolved at the same short-term scale. The scale problems are associated with intrusion of mixed water, sub-surface maxima of plankton growth and the time scale of vertical mixing, see e.g., Pedersen [1993] and Højerslev [1985]. In chapter 5, examples of such short-term and local-scale problems are described.

Monitoring

The monitoring of marine systems must take the hydrographic variability of the system into account. The time variation in biological parameters as illustrated by the spring bloom, and the time variation in the hydrographic parameters as illustrated by the intermittence of the flow through the Belts, show how the monitoring planning must take the time variation into account.

Monitoring of coastal systems is expensive. It is therefore crucial to define the objective of the monitoring programmes before commencing the monitoring. Poor planning can result in oversampling (waste of money) or, even worse, in incomplete evaluation of questions [National Research Council, 1990]. Models of the system variability and structure can assist in planning the monitoring. As an example, Wulff and Rahm [1989] have used modeling to demonstrate that only a few well chosen stations with a high sampling rate are necessary to monitor the nutrient conditions in the Baltic Sea, as opposed to many but seldomly visited stations.

Modeling

When the physical description of the hydrography and its relation to the biological processes are considered quantitatively, mathematical models have to be used. Models of the hydrographic processes and the general circulation have long been used for special purposes, examples from the Danish Straits are the forecast model for currents and water level by Vested et al. [1991] and the circulation model by Stigebrandt [1983].

When addressing the applicability of models in the context of eutrophication it should be born in mind that a numerical model is nothing more than a simplified theoretical abstraction of a natural system. Elements that are considered to be important for the task at hand are expressed in standard mathematical terms, and calculations are performed by computer. Numerical models are used to synthesize information, answer questions, solve problems or make decisions. Examples of models that combine the hydrographic and biological systems are given in chapter 11.

Conclusions

The Danish Straits are probably some of the best studied regions of the world. The hydrography of the Danish Straits has been studied and monitored for more than a century. The knowledge of the hydrographic processes and its interaction with biology have reached a stage where lack of measurements is hardly the reason for lack of progress in science. The future development in knowledge of this interesting system will be achieved through the combined use of numerical modeling, remote sensing and detailed measurement programs. As it has been pointed out in the previous sections, the variability and patchiness of the biological parameters are closely linked to the variable hydrographic conditions. Further progress in the field of biology is therefore closely linked to the understanding of the relation between the hydrographic and biological processes.

Of particular interest for future study is the vertical mixing processes and their importance for the plankton dynamics. Much has to be learned before the combination of vertical and horizontal transport processes is sufficiently understood and reduced to a level that can be modeled.

The monitoring strategies have until now been based mainly on traditional vessel-based cruises. This monitoring method will be transformed in the future when remote sensing, automated (on-line) monitoring and modeling tools become more widespread.

References

Andersson, L., and L. Rydberg, Exchange of water and nutrients between the Skagerrak and the Kattegat, *Estuar. Coast. Shelf Sci.*, *36*, pp. 159–181, 1993.

Fenchel, T., (Ed.), *Plankton Dynamics and Carbon and Nutrient Flow in Kattegat* (in Danish), *Havforskning fra Miljøstyrelsen*, *10*, 240 pp., Danish Environmental Protection Agency, Copenhagen, 1992.

Hansen, I. S., G. Ærtebjerg, L. A. Jørgensen and F. B. Pedersen, Analysis of the Oxygen Depletion in the Kattegat, the Belt Sea and the Western Baltic Sea (in Danish), *Havforskning fra Miljøstyrelsen*, *1*, 133 pp., Danish Environmental Protection Agency, Copenhagen, 1990.

Hansen, I. S., G. Ærtebjerg, K. Richardson, J. Heilman, O. V. Olesen, and F. B. Pedersen, The effect of reduced nitrogen emissions on oxygen conditions in inner Danish waters (in Danish), *Havforskning fra Miljøstyrelsen*, *29*, 103 pp., Danish Environmental Protection Agency, Copenhagen, 1994.

Højerslev, N. K., Bio-optical measurements in the Southwest Florida Shelf ecosystem, *J. Cons. Intl Explor. Mer, 42*, 65–82. 1985.

Jacobsen, T. S., *Sea Water Exchange of the Baltic: Measurements and Methods, The Belt Project*, Ministry of the Environment, Copenhagen, 1980.

Jakobsen, F., The major Inflow to the Baltic Sea during January 1993, *J. Mar. Systems., 6(3)*, 227–240, 1995.

Jakobsen, F., G. Ærtebjerg, N. K. Højerslev, N. Holt, J. Heilman and K. Richardson, Hydrographic and biological description of the Skagerrak front (in Danish), *Havforskning fra Miljøstyrelsen, 49*, 106 pp., Danish Environmental Protection Agency, Copenhagen, 1994.

Ministry of the Environment, *The Belt Project* (in Danish), Ministry of the Environment, Denmark, 1976.

Møller, J. S., and I. S. Hansen, Hydrographic Processes and Changes in the Baltic Sea, *Dana, 10*, 87–104, 1994.

Møller, J. S. and C. B. Pedersen, Analysis of hydrographic data from southern Kattegat (in Danish), *Havforskning fra Miljøstyrelsen, 20*, 114 pp., Danish Environmental Protection Agency, Copenhagen, 1993.

National Research Council, *Managing Troubled Waters, the Role of Marine Environmental Monitoring,* by Committee on a System Assessment of Marine Environmental Monitoring, Marine Board, Commission on Engineering and Technical Systems, National Research Council. National Academy Press, Washington, D.C., 1990.

Pedersen, F. B., Fronts in the Kattegat: the hydrodynamic regulating factor for biology, *Estuaries, 16*, 104–112, 1993.

Pedersen, F. B., Environmental hydraulics: stratified flows, *Lecture Notes on Coastal and Estuarine Studies, 18*, Springer, 1986.

Pedersen, F. B., The oceanographic and biological tidal cycle succession in shallow sea fronts in the North Sea and the English Channel, *Estuar. Coast. Shelf Sci., 38*, 1994.

Poulsen O., Dynamics of the Skagerrak-front (in Danish), *Havforskning fra Miljøstyrelsen, 7*, Danish Environmental Protection Agency, Copenhagen, 1991.

Richardson, K., Algal blooms in the North Sea: The good, the bad and the ugly, *Dana, 8*, 83–93, 1989.

Rodi, W., Mathematical modeling of turbulence in estuaries, in *Lecture Notes on Coastal and Estuarine Studies, 1: Mathematical Modeling of Estuarine Physics*, edited by J. Sündermann and K.-P. Holz, pp. 14–26, Springer, 1980.

Rodi, W., *Elements of the Theory of Turbulence, in Coastal, Estuarial and Harbour Engineers Reference Book*, edited by M. B. Abbot and W. A. Price. E. & F. N. Spon, London, 1994.

Stigebrandt, A., A model for the exchange of water and salt between the Baltic and the Skagerrak, *J. Phys. Oceanogr. 13*, 411–427, 1983.

UNESCO, Background papers and supporting data on the International Equation of Seawater 1980, *Technical Papers in Marine Science 38*, 1981 (a).

UNESCO, Background papers and supporting data on the International Equation of Seawater 1980, *Technical Papers in Marine Science 37*, 1981 (b).

Ünlüata, T. Oguz, M. A. Latif and E. Özsoy, On the physical oceanography of the Turkish straits, in *The Physical Oceanography of Sea Straits*, edited by L. Pratt, pp. 25–60. Kluwer, Dordrecht, The Netherlands, 1990.

Vested, H. J., H. R. Jensen, H. M. Pedersen, A.-M. Jørgensen and B. Machenhauer, An operational hydrographic warning system for the North Sea and the Danish Belts, *Continental Shelf Res., 12*, 65–81, 1992.

Wulff, F., and L. Rahm, Optimising the Baltic sampling programme: the effects of using different stations in calculations of total amount of nutrients, *Beitr. Meereskunde, Berlin, 60*, 61–66, 1989.

Weidemann, H., Untersuchungen über unperiodische und periodische hydrographische Vorgänge in der Beltsee, *Kieler Meeresforsch., 7*, 70-86, 1950.

4

Material Flux in the Water Column

Thomas Kiørboe

4.1. Introduction

There is a continuous exchange of substances between the pools of solute and particulate material in the pelagic zone. For example, inorganic solute nutrients and carbon dioxide are incorporated into particulate organic matter by photosynthesis and are subsequently remineralized by metabolic processes. Other quantitatively important processes are summarized in Figure 4.1. Mechanisms of transport of solute and particulate substances differ radically. While both particles and solutes may be transported with the surrounding water and undertake Brownian movements (the latter only important for solute molecules and very small particles, however) particles are in addition subject to gravity and, in the case of organisms, may

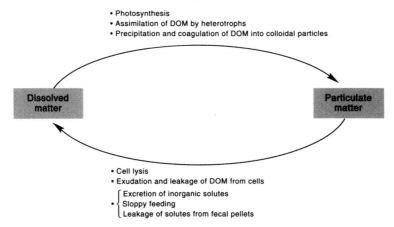

Figure 4.1. Exchange of substance between the pools of dissolved and particulate matter. Summary of quantitatively important processes.

Eutrophication in Coastal Marine Ecosystems
Coastal and Estuarine Studies, Volume 52, Pages 67–94
Copyright 1996 by the American Geophysical Union

TABLE 4.1. Mechanisms of transport of particles and solutes

Transport mechanism	Solutes	Particles
Brownian motion (diffusion)	+	Important for particles < 1 μm
Advection	+	+
Gravity	−	Important for particles > 1 μm and w. density > medium
Motility	−	Important only for swimming organisms

move actively (Table 4.1). This has several implications for the exchange of substance between the various pools in the pelagic zone and for the distribution of matter. One consequence is that solute and particulate material, as well as the different processes associated with particles and solutes, are differently distributed in the vertical. For example, photosynthesis occurs only in the euphotic zone due to its dependency on light, while sedimenting phytoplankters and organic particles to a large extent are remineralized below the euphotic zone. Stratification of the water column and the vertical displacement of these oxygen-generating and -consuming processes may cause anoxia to develop below the pycnocline.

The purposes of this chapter are to

1. describe the different mechanisms and properties of the exchange of substance between the solute and particulate pools in the water column;
2. describe the different mechanisms of transport of solute and particulate material in the pelagic zone and
3. to discuss the implications of the exchange and transport processes for pelagic food web structure and for the supply of oxygen-consuming material to the sea floor in coastal marine waters. Throughout this chapter I will use conceptual examples, preferentially from Danish waters, to illustrate the discussion. A subsequent chapter (Richardson, chapter 5) will provide flux budgets for the Kattegat specifically.

4.2. Exchange of Substances between the Dissolved and Particulate Pools

Table 4.2 gives an overview of the types, sizes and typical concentrations of particulate organic material in coastal waters. The term "particle" is here used in its widest sense and includes also living organisms. Solutes may be transformed to particulate material by at least three different processes (Figure 4.1): i) uptake of inorganic nutrients by phytoplankton and bacterioplankton, ii) uptake of dissolved organic matter, primarily by heterotrophic microorganisms; and iii) precipitation and coagulation of solute organics to colloidal particles. These processes are discussed separately in the following. The other types of particles listed in Table 4.2 are the result of trophic interactions (zooplankton) or physical processes (aggregates). Organic particles may be remineralized to solute inorganics by heterotrophs and solute organics may be exuded or leak from living cells or organic particles. These processes will be dealt with subsequently.

TABLE 4.2. Types and typical sizes and concentrations of organic particles occurring in coastal waters

Type	Size, cm	Typical conc., # ml^{-1}	Reference
Colloids	$5-120 \times 10^{-7}$	10^9	Wells & Goldberg 1991
Vira	5×10^{-6}	10^7	Børsheim 1992
Submicron particles	$0.4-0.9 \times 10^{-4}$	10^7-10^8	Longhurst et al. 1992; Koike et al. 1990
Bacteria	$0.3-1.0 \times 10^{-4}$	10^5-10^6	Fogg 1986
Transparent exopolymeric particles (TEP)	$5-50 \times 10^{-4}$	10^2-10^4	Alldredge et al. 1993
Marine snow	$10^{-1}-10^1$	$10^{-3}-10^{-2}$	Alldredge & Silver 1988
Picoalgae	$0.2-2 \times 10^{-4}$	10^4	Fogg 1986
Nanoalgae	$2-20 \times 10^{-4}$	10^3	"
Microalgae	$2-20 \times 10^{-3}$	10^2-10^3	"
Microzooplankton	$2-100 \times 10^{-4}$	$10^{-2}-10^0$	"
Mesozooplankton	$2-20 \times 10^{-1}$	$10^{-3}-10^{-4.}$	"

4.3. Conversion of Dissolved Inorganic Material to Organic: Primary Production

Primary production of the phytoplankton lies at the basis of pelagic food webs and provides the most important energy input for heterotrophic processes in the water column and on the sea floor. The accumulation of phytoplankton biomass in the euphotic zone is limited by light during winter and in deeply mixed water columns, but is normally assumed to be limited by the availability of inorganic nutrients during the productive summer season. When the nutrient concentration is low the uptake rate of nutrients (U) by a spherical phytoplankton cell depends on the concentration of nutrients in the ambient water (C), the size of the cell (characterized by its radius, r), the molecular diffusivity (quantified by the diffusion coefficient, D) and on the advective transport of nutrients to the cell surface (characterized by the dimensionless Sherwood number, Sh) (see e.g. Berg and Purcell [1977] for a formal proof):

$$U = 4\pi r D Sh C \tag{1}$$

The Sherwood number is the ratio of total nutrient transport to the cell surface (by advection and diffusion) to the transport by diffusion. If the advective transport is zero, $Sh = 1$, and the uptake rate is governed only by the rate of molecular diffusion. From Eq. 1 it follows that in the diffusion-limited case (i.e., $Sh = 1$) the mass (or volume) specific uptake rate (U/cell volume $= (4\pi r D Sh C)/(\frac{4}{3}\pi r^3) = 3D Sh C/r^2$) de-

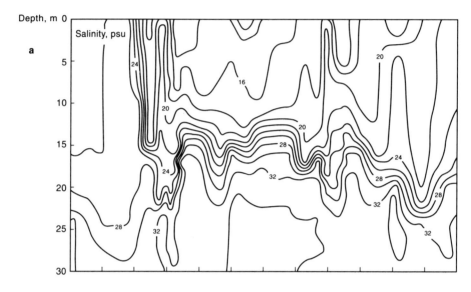

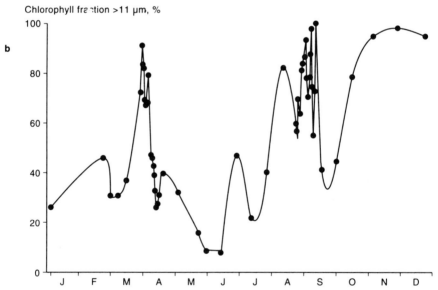

Figure 4.2. Seasonal variation in vertical water column structure (a) and size composition of the phytoplankton (b) in southern Kattegat. In the Kattegat vertical water column structure is determined mainly by salinity. Blooms of phytoplankton occur in association with the vernal stratification of the water column (March–April), beginning deepening of the upper mixed layer in late summer (August–September), and in late autumn (November–December) where the water column becomes deeply mixed. These blooms are all dominated by phytoplankton retained on an 11-μm screen, while cells < 11 μm characterize the phytoplankton during summer stratification. Data from Kiørboe and Nielsen [1994].

clines with the square of the cell radius. Thus, small cells are very much more efficient than large cells in collecting inorganic nutrients from the environment. For example, a 1-μm cell extracts nutrients 100 times more efficient than a 10-μm cell. Or more generally, picophytoplankton (0.2–2.0 μm) are much more efficient than nanophytoplankton (2.0–20 μm) that, in turn, are much more efficient than microphytoplankton (> 20 μm) cells in the competition for inorganic nutrients.

In spite of this, not all phytoplankton cells in the ocean belong to the smallest size class. Empirical evidence suggests that microphytoplankton blooms and dominates the phytoplankton biomass in localized or episodic, nutrient-rich, turbulent environments while very small or motile forms dominate in oligotrophic, stagnant waters (see Kiørboe [1993] for a review). Turbulent mixing brings inorganic nutrients toward the surface while stratification retains the phytoplankton in the euphotic zone. Transitions in time or space in the vertical water column structure are, therefore, indicative of ephemeral or localized enhancements of phytoplankton growth conditions, and this is where the microphytoplankton blooms and dominates the plankton flora. Some examples from Danish waters are shown in Figures 4.2 and 4.9. Apparently, the competitive disadvantage of large cell size is offset by other factors in some environments. One such factor is the advective transport of nutrients to the cell surface (i.e., $Sh > 1$). In the diffusion-limited case (i.e., when the uptake of nutrients across the cell surface exceeds the delivery by diffusion) a nutrient-depleted region around the cell will be established. If the delivery of molecules to this region is enhanced by advection, either because the cell swims or sinks through the water or because of local fluid velocity gradients (shear, turbulence), then the uptake rate will increase. The magnitude of increase is given by the Sherwood number – the Sherwood number is the factor with which diffusion transport is increased due to fluid motion. It is not straightforward to calculate the Sherwood number (for examples see e.g. Logan and Alldredge [1989], Kiørboe [1993], Murray and Jackson [1993]); it depends on whether the cell sinks/swims or is suspended in a turbulent shear field and it is a function of the cell size, the shear rate and the swimming /sinking rate. Table 4.3 gives some estimates of Sherwood numbers for differently

TABLE 4.3. Sherwood numbers calculated for settling (Sh_{sett}) and swimming (Sh_{swim}) cells and for cells suspended in a turbulent fluid shear field (Sh_{turb}). The Sherwood number is the factor by which nutrient uptake in cells is increased due to advective transport. Calculations of Sherwood numbers from equations presented in Kiørboe [1993]. Settling rates (v, cm s^{-1}) calculated from cell radius (r, μm) as $v = 2.48 r^{1.17}$ [Jackson, 1990]. Swimming velocity (v, cm s^{-1}) calculated for flagellates from their equivalent spherical diameter (ESD, μm) as $v = 9.3 \times 10^{-2} ESD^{0.26}$ [Sommer 1988]. For calculation of Sherwood number in a turbulent flow field a shear rate of 1 s^{-1} has been assumed; this is equivalent to the average turbulent shear generated in a 10-m upper mixed layer by a wind of ca. 6 m s^{-1} [MacKenzie and Leggett, 1991].

Cell diameter, μm	Sh_{sett}	Sh_{swim}	Sh_{turb}
1	1.007	1.84	1.004
10	1.034	5.41	1.041
100	1.116	23.1	1.41
1000	11.24	–	5.11

sized cells that are swimming, sinking or subject to fluid motion. Small cells gain very little from moving through the water (swimming, sinking) or from fluid motion (turbulence), while the nutrient uptake in large cells is significantly increased. Thus, the competitive pressure for small size is relaxed in a turbulent environment and for swimming cells (flagellates), and this is consistent with the observed distributions of the various forms. However, the effect of the advective transport in large cells is insufficient to entirely compensate for the declining size-specific nutrient uptake by molecular diffusion with increasing cell size.

Another factor that may help offset the competitive disadvantage of being large and, thus, explain the occurrence of large phytoplankters in episodic, nutrient-rich environments is predation [Munk and Riley, 1952, Geider et al., 1986, Kiørboe, 1991]. The growth of phytoplankton cells is, unlike the growth of their zooplankton predators, only weakly related to cell size [e.g., Banse, 1982b; Blasco et al., 1982, Nielsen and Sand-Jensen, 1990]. Small phytoplankton cells are preyed upon by small predators, e.g. heterotrophic nanoflagellates, and the phytoplankton prey and their predators have similar population growth rates (on the order of hours). Thus, the predators of small phytoplankton cells have the capacity to control their phytoplankton prey populations. Large phytoplankton cells are eaten by larger metazoans, e.g. copepods, and these have generation times one to several orders of magnitude longer than those of the prey. Thus, the numerical response in predator populations to variations in phytoplankton prey population sizes will be increasingly delayed with increasing phytoplankton cell size.

When nutrients are episodically injected into the euphotic zone, population sizes of both small and large phytoplankton cells will start to increase. However, populations of small phytoplankton cells will soon be caught up by their predators, while the large cells may continue their growth largely unutilized by grazers until the inorganic nutrients have been exhausted. This may explain why microphytoplankton dominates and blooms in episodically nutrient-enriched environments.

The interplay between nutrients and predators in controlling the phytoplankton biomass may help explain an apparent paradox. As noted above biomass accumulation in the water column is normally ultimately limited by nutrient availability. At the same time it appears that the growth rate of the phytoplankton is rarely limited by nutrient availability; rather, the phytoplankton grow at maximum rates limited only by the availability of light [e.g., Goldman, 1987]. Thus, phytoplankton biomass is nutrient-limited while its specific growth rate is light-limited. An example of this apparent paradox is provided by the seasonal variation in nutrient concentration, phytoplankton biomass, specific growth rate of the phytoplankton, and daily irradiance in the Kattegat (Figure 4.3). Phytoplankton biomass accumulates primarily during the vernal water column stratification when the concentration of inorganic nutrients is high, and declines to low values when the nutrients have been exhausted. Phytoplankton growth rate, on the other hand, is highest during summer stratification when the nutrient concentration is below the detection limit, and rather follows the seasonal variation in irradiance and, hence, availability of light. Nutrient limitation of phytoplankton growth is likely to occur only subsequent to episodic fertilization of the euphotic

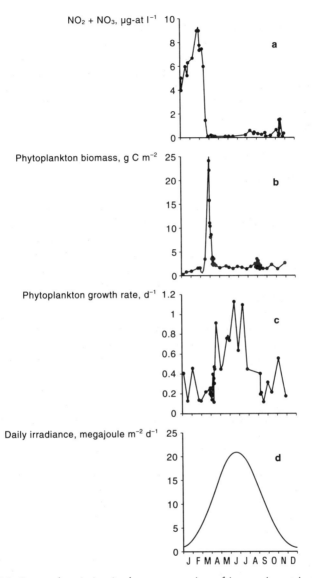

Figure 4.3. Seasonal variation in the concentration of inorganic nutrients (NO_2 + NO_3) in the upper mixed layer (a), phytoplankton biomass (b), phytoplankton growth rate (c), and daily irradiance (d) in southern Kattegat. Phytoplankton biomass depends on the concentration of inorganic nutrients (note that the nutrient scale is logarithmic), while phytoplankton growth rate appears to be independent on nutrient concentration but to vary in proportion to irradiance. Phytoplankton biomass was estimated from depth-integrated chlorophyll by assuming a chlorophyll:carbon ratio of 50. The phytoplankton growth rate was calculated as the estimated daily primary production (^{14}C incubations) divided by phytoplankton biomass. Daily irradiance is a standard curve for the location. Data from Kiørboe and Olsen (unpublished) and from Richardson and Christoffersen [1991].

zone, where the phytoplankton may "escape" predator control and exhaust the nutrient source. In oligotrophic waters, large cells are replaced by smaller forms due to competition for nutrients and these are controlled by predation. Hence, nutrients are regenerated at the same rate as they are being utilized by the phytoplankton, and the turnover rate is determined by the light-limited growth rate of the phytoplankton. Because the phytoplankton production rate (primary production) is the product of biomass and growth rate it is limited by either light and/or inorganic nutrients.

4.4. Heterotrophic Uptake of Dissolved Organic Matter: Bacterial Production

Dissolved organic matter may be assimilated by cells and thereby transformed to particulate organic matter. Uptake of dissolved organic matter is, of course, limited by the same physical constraints as the uptake of dissolved inorganics, and the uptake rate can, thus, be described by Eq. 1. The concentration of dissolved organic matter (DOM) in coastal regions is relatively high (typically 1–3 mg C l^{-1}), but the majority (80–90%) is normally considered to be very refractory and to have long turnover times [Søndergaard and Middelboe, 1995]. It should also be noted that the dissolved fraction is operationally defined as material passing an 0.2-μm filter and some of the "solute" DOM may in fact be particulate and occur as colloidal or submicron particles (see below); recent estimates suggest that > 10% [Koike et al., 1990], > 30% [Benner et al., 1992] or about 16% [Kepkay et al., 1993] of the material passing an 0.2-μm filter is particulate. Thus, the concentration of small organic molecules that can be assimilated by cells is low, even in coastal regions, and approximately equivalent to the carbon biomass of pelagic bacteria. Because of that and due to the constraints implied by Eq. 1, only very small cells can efficiently utilize the source of dissolved organics. Heterotrophic bacteria dominate the picoplankton and are, consequently, the main consumers of DOM.

The quantitative importance of bacterial production as a source of particulate material in coastal regions is significant. Particle production due to bacterial assimilation and utilization of DOM equals on the average some 30% of the particulate primary production [Cole et al., 1988]. In some cases bacterial production may approach or even exceed the primary production [e.g., Børsheim, 1992].

Taking into consideration the fact that all dissolved organics ultimately stem from phytoplankton production in a pelagic food web without external input and given the growth yield of pelagic bacteria (0.1–0.6 [Børsheim, 1992]), it may at first appear paradoxical that bacterial assimilation is so high relative to primary production. It is a frequent misunderstanding, however, that heterotrophic assimilation and secondary production (i.e., the sum of production of all heterotrophic organisms) in a closed system cannot exceed the primary production [Strayer, 1988]. This can easily occur if the organic matter passes several consumers before eventually being respired. Note that the energetic constraint is that total *respiration* cannot exceed primary production. Thus, a high ratio of bacterial to primary production implies that dissolved organics are taken up and excreted by bacteria – or bacterial predators – several times; i.e., that DOM is recycled several times. This is likely to be the case also in neritic waters, even though allochtonous organic matter from river runoff may contribute to bacterial production. In coastal waters, the flux of labile terrigenous organic matter corresponds to only a few percent of the primary production; even in estuarine waters terrigenous organic matter flux is < 20% of the primary production [Smith and Hollibaugh, 1993].

4.5. Precipitation and Coagulation of Dissolved Organic Material

Another potentially important but as yet incompletely understood mechanism of particle formation from solute substances is by physico-chemical processes. Recently, Wells and Goldberg [1991] described the occurrence of small (5–120 nm) colloidal particles in high concentrations ($\approx 10^9$ ml^{-1}) in waters off California (Table 4.2). The organic nature and the vertical stratification of these particles suggest them to be of biotic origin, although their mode of formation is uncertain. Koike et al. [1990] and Longhurst et al. [1992] found concentrations on the order of 10^7–10^8 ml^{-1} of organic submicrometer particles (0.4–1.0 µm) in surface waters of the north Pacific and off Nova Scotia, respectively. This size range would include bacteria and picophytoplankton, but these organisms contributed < 5% of the particles and the majority were, thus, nonliving [Koike et al., 1990]. One potential mode of formation of submicrometer particles is by nanoflagellate-bacterial interaction [Koike et al., 1990]. Finally, Alldredge et al. [1993] discovered significant concentrations (up to 5000 ml^{-1}) of larger (2–100 µm) transparent exopolymeric particles (TEP) in surface waters off California. Even larger mucus particles, or aggregates consisting of mucus and solid particles, several tens of cm in diameter, have been described from the Adriatic [Stachowitsch et al., 1990]. Both the particles of TEP and the giant mucus particles in the Adriatic can be formed from polymeric material exuded by diatoms [Stachowitsch et al., 1990; Kiørboe and Hansen, 1993, Passow et al., 1994], bacteria, and from macrophyte leachates [Alber and Valiela, 1994]. Alldredge et al. [1993] hypothesized that colloidal particles form from exopolymeric solutes by cation bridging of 3–10 nm microfibrils, that these further coagulate or otherwise coalesce to colloidal particles, as has been described in fresh water [Jensen and Søndergaard, 1982], and subsequently to the larger particles of TEP. Thus, particulate material may form directly from DOM by physico-chemical processes, and the concentrations recorded for these various types of organic particles (Table 4.2) suggest that this may be a major pathway from solutes to particles. These different types of recently discovered organic particles may have important implications for transport processes in the plankton; they may act as sorption sites for solutes [Moral and Gschwend, 1987; Wells and Goldberg, 1991] and attachment sites for bacteria [Alldredge et al., 1993; Alber and Valiela, 1994], serve as food for particle grazers [Shimeta, 1993], and coagulate with other particles, including phytoplankton [Passow et al., 1994; Kiørboe and Hansen, 1993], into marine snow aggregates and thus enhance the vertical transport of substances by sedimentation (see below).

4.6. The Transformation of Particulate to Dissolved Matter

Matter bound in suspended particles may be transformed to the solute state, either in the form of inorganic ions, or in the form of dissolved organic matter. The generation of inorganic nutrients is the inescapable result of metabolic processes, by which organic substances are oxidized to their low-energy inorganic components. These will be excreted to the environment by the heterotrophic organisms. Scaling arguments suggest that, on average, the majority of the remineralization will be due to very small organisms: because the biomass of planktonic organisms is on average approximately constant in equal, logarithmic size classes [Sheldon et al., 1972], and because the weight-specific metabolic rate of organisms declines with the body mass raised to an exponent of ca. -0.25 [e.g., Banse 1982a], organisms smaller than 100 μm account for > 90% of total pelagic remineralization.

Due to the significance of small organisms and because small particles sink very slowly (see below), the majority of the organic matter produced in the euphotic zone will be remineralized within the zone if the particle size distribution fits the equilibrium Sheldon-distribution. However, if the size distribution is pertubated toward larger particle sizes due, for example, to nutrient enrichment and/or aggregation of particles (see below) sedimentation of particulate material may displace a significant fraction of the remineralization process to below the pycnocline.

Dissolved organic compounds are excreted or lost directly or indirectly from organisms and ultimately stem primarily from the phytoplankton; river runoff contributes only little additional organic matter, even in coastal waters (cf. above). The "demand" for dissolved organic matter in the water column is high. The requirements for bacterial uptake of DOM and for the formation of colloidal and larger nonliving organic particles directly from DOM together may exceed the pelagic primary production. Bacterial production averages 30% of primary production in aquatic environments [Cole et al., 1988]. With an assumed growth efficiency of 40% [Bjørnsen and Kuparinen, 1991], bacteria would alone on average require the equivalence of 75% of the primary production as DOM. We have no quantitative estimates of the turnover rate of colloidal organic material. However, carbon concentrations of such particles are similar to carbon concentrations of the phytoplankton (on the order of 100 μg C l^{-1}, Wells and Goldberg, 1991) and theoretical [Shimeta, 1993] as well as experimental studies [e.g., González and Suttle, 1993; Tranvik et al., 1993; Marchant and Scott, 1993] suggest that these particles are ingested by nanoflagellates at high rates. It follows that the turnover of colloidal organic particles and, hence, the demand for DOM is relatively high. Since most DOM in the water column ultimately stems from photosynthetic carbon fixation, these considerations suggest that recycling of organic matter is quantitatively very important.

This conclusion is further substantiated by the fact that direct exudation of DOM from phytoplankton cells is presumably small. While the older estimates of DOM leakage suggested that up to > 70% the primary production can be lost in solute form [e.g., Fogg, 1983], more recent estimates suggest that realistic exudation rates are rather on the order of 10% of primary production [Baines and Pace, 1990], or lower (see Jumars et al. [1989]). The lower estimates are also consistent with the theoretical considerations of Bjørnsen [1988]. Bjørnsen suggested that exudation of DOM from healthy phytoplankton is due primarily to passive diffusion across the permeable cell membrane. Accordingly, the exudation rate depends on the cell surface area. Bjørnsen's model combined with realistic assumptions of phytoplankton cell membrane permeabilities leads to phytoplankton exudation rates that are comparable with those measured experimentally, i.e., 1–10% of the primary production (see also Kiørboe [1993]). However, because the exudation rate depends on the cell surface area, and because bacterial cell surface area per unit volume of sea water exceeds the cell surface area of all other organisms combined [Williams, 1981], these considerations would also suggest that leakage of DOM from bacteria would be an important source of dissolved organic matter. Thus, a significant fraction of organic matter recycling in the water column may occur within the microbial community.

Jumars et al. [1989] likewise maintained that recycled DOM is a more important source than direct leakage from phytoplankton cells. These authors emphasized the importance of inefficient grazing and digestion: DOM may result from "sloppy feeding", i.e., spillage of dissolved compounds from break-up of phytoplankton cells during the digestion process, or from leakage of solutes from fecal pellets. By simple diffusion-advection models, they showed that solutes contained in fecal pellets would be rapidly leaked to the environment and, thus, remain in the water column rather than falling out with the pellet itself. Jumars et al. [1989] estimated that these pathways could account for a DOM flux equivalent to 20–70% of the primary production, several fold higher than the leakage rate of DOM directly from phytoplankton cells (see also Baines and Pace [1990]).

4.7. Mechanisms of Transport of Solute Inorganic Nutrients

The foregoing discussion of exchange processes between the solute and particulate pools of material in the water column has also considered small-scale diffusion-advection transport of solutes; i.e., transport in and out of particles. I will now consider the larger-scale processes and focus on the transport of inorganic nutrients: What are the sources of inorganic nutrients, and how are they transported to the euphotic zone where they can be utilized by the phytoplankton?

There are two primary sources of inorganic nutrients for primary production in the euphotic zone, viz. "new" and "regenerated" nutrients [Dugdale and Goering, 1967]. Regenerated nutrients are due to remineralization of organic material in the euphotic zone while new nutrients are transported to the euphotic zone from elsewhere. New nutrients may be transported to the euphotic zone by various mechanisms; viz. by atmospheric transport and subsequent dry or wet deposition on the sea surface (e.g. Larsen and Assam, chapter 2, this volume), land runoff (Møller, chapter 3, this volume), upwelling, vertical mixing, entrainment across the pycno-

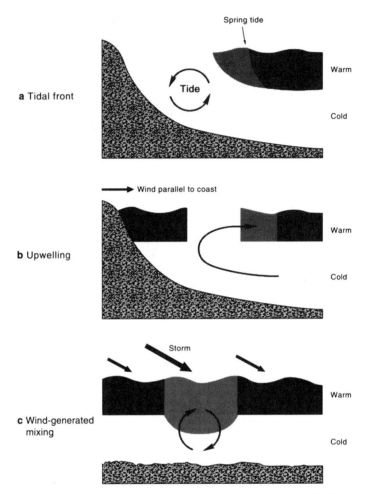

Figure 4.4. Schematic of various hydrodynamic processes that may enrich the euphotic zone with inorganic nutrients. Intensity of hatching from white to grey indicate variation from cold nutrient-rich to warm nutrient-poor water. a: The position of a tidal front varies with the neap-spring tidal cycle. Thus, in a fortnightly cycle cold nutrient-rich water stratifies, and phytoplankton is withheld in the euphotic zone, where they bloom. b: Alongshore winds cause deep, nutrient-rich water to ascend near the coast. When the upwelled water stratifies, the phytoplankton will bloom. c: Episodic wind events may cause a temporary deepening of the upper mixed layer and, thus, mixing of nutrient-rich cold water into the nutrient-poor surface layer; this will cause the phytoplankton to bloom.

cline, cross-frontal advection of baroclinic eddies, etc. Several hydrodynamic processes of nutrient enrichment of the euphotic zone have been schematically depicted in Figure 4.4. While atmospheric deposition of inorganic ions, diffusion of nutrients across the pycnocline as well as local remineralization of nutrients may be more or less continuous processes, most hydromechanical processes give rise to episodic injections of nutrients into the euphotic zone. For example, wind events may temporarily deepen the upper mixed layer of the water column, and thus make nutrients immediately below the pycnocline available to surface populations of the phytoplankton. Likewise, the intensity of coastal upwelling depends on the strength of alongshore winds and is, therefore, episodic. Also, entrainment of deep, nutrient-rich water into the upper mixed layer is driven by velocity differences between the two layers, and this will depend on the intensity of the tidal or wind-generated currents. Horizontal excursion of tidal fronts will follow the neap-spring tidal cycle and depends also on the wind intensity. Thus, winds, and currents generated by winds, may cause episodic injections of nutrients into the euphotic zone.

As noted above the level of phytoplankton biomass is typically directly dependent on the availability of inorganic nutrients. However, the time scale of the nutrient-enrichment processes has implications for the size composition of the developing phytoplankton community. If the nutrient supply (whether new or regenerated) is continuous, a phytoplankton community dominated by small forms is expected to develop, while episodic availability will lead to dominance of larger forms. The larger forms will typically bloom at such spatio-temporal oceanographic "singularities" [Legendre and Le Févre, 1989] where nutrients are episodically available. As discussed above, this is mainly due to the different capability of the predators of large and small phytoplankton cells to control the population sizes of their prey.

4.8. Mechanisms of Transport of Suspended Particulate Matter

Suspended particles may move with the surrounding water and settle by gravity. In this section we shall in particular consider the vertical flux of particulate material. Sedimentation of organic particulate material may relocate the oxygen-consuming degradation of particulate organic material (POM) from the euphotic zone to below the pycnocline. If the sedimentation of POM, measured in reduction equivalents, exceeds the downward mixing and entrainment of oxygen, anoxia will eventually develop. The amount of sedimenting POM depends, of course, on the biomass of suspended POM which, in turn, depends on the availability of inorganic nutrients. However, if the residence time of suspended POM in the euphotic zone is long, the organic particles may be eaten or otherwise degraded in the upper mixed layer. The vertical flux of particulate material, therefore, also depends on the sinking velocity of suspended POM and, thus, on the processes that affects sinking velocity. I shall consider these processes in some detail in the following.

The fall velocity (v) of a suspended particle depends on its excess density ($\rho - \rho'$) and on the gravitational acceleration (g) on one hand and on the frictional resistance of the water on the other; the latter is a function of the size of the particle and of the viscosity (η) of the water. The fall velocity of a spherical particle of radius, r, is described by Stokes' law:

$$v = \tfrac{2}{9}\, g r^2 (\rho - \rho')\eta^{-1} \tag{2}$$

TABLE 4.4. Settling velocities for spherical phytoplankton cells of different sizes calculated from Stokes' law. In (1) a constant differential cell density of 0.05 g cm^{-3} has been assumed; i.e $v = 1091r^2$ cm s^{-1}. In (2) the declining cell density with increasing cell size has been taken into account; i.e., $v = 2.48r^{1.17}$ cm s^{-1} [Jackson, 1990]. From Kiørboe [1990].

Cell diameter, μm	Settling velocity, m d^{-1}	
	(1)	(2)
1	2.36×10^{-3}	1.99×10^{-2}
10	2.36×10^{-1}	2.94×10^{-1}
100	2.36×10^{1}	4.35×10^{0}
1000	2.36×10^{3}	6.44×10^{1}

In Table 4.4 Stokes' settling velocities for four size categories of spherical phytoplankton have been calculated; cells smaller than ca. 100 μm in diameter sink insignificantly, while larger cells sediment rapidly out of the upper mixed layer.

Observed density and size distributions of suspended particles are frequently insufficient to account for the observed vertical flux of particulate matter. For example, sinking of phytoplankton subsequent to blooms may exceed the predicted Stokes' settling velocity by orders of magnitude [e.g., Smetacek, 1985]. Two examples from Danish waters may serve to illustrate this (I shall subsequently return to these examples):

1. During an early spring bloom of the diatom *Skeletonema costatum* in the Kattegat, Olesen [1993] observed the cells to sink with velocities of up to near 4 m d^{-1}. *S. costatum* cells measure approximately 5 μm in diameter; thus, according to Table 4.4, they are expected to sink 1–2 orders of magnitude slower. However, *S. costatum* is chain-forming and possesses spines, and this may change its fall velocity. It is therefore more relevant to compare the observed sinking velocity to experimental determinations of *S. costatum* fall velocities; these range between 0.1–1.4 m d^{-1} [Smayda and Boleyn, 1966; Riebesell, 1989]. The sinking velocity of *S. costatum* observed in the Kattegat and elsewhere (e.g. 40–50 m d^{-1} during a spring bloom in the Baltic, [von Bodungen et al., 1981]), thus, significantly exceeds those determined experimentally.
2. Throughout summer in the Kattegat, Olesen and Lundsgaard [1994] found a relatively high and nearly constant sedimentation of particulate organic material to the sea floor, ca. 300 mg C m^{-2} d^{-1}. Since the concentration of organic particles, excluding zooplankton, is also relatively constant and on the order of 150 mg C m^{-3} during the same period, an average sinking velocity of suspended organic particles of ca. (300 mg C m^{-2} d^{-1})/(150 mg C m^{-3}) = 2 m d^{-1} can be calculated. During summer the phytoplankton community is dominated by nanoflagellates [Thomsen, 1992] and the size distribution of suspended particles obtained by electronic particle counter shows that the majority of the particle-biomass in the size range 2–200 μm is in particles < 10 μm (own observation). Such small particles will sink < 0.25 m d^{-1}. Thus, here again we find a significant difference between expected fall velocity and bulk POM settling velocities.

The reason for these discrepancies between observed and expected settling velocities is that suspended particles may aggregate into large (mm- to cm-size), rapidly sinking "marine snow" particles. When small particles combine into larger aggregates, their fall velocity will increase according to Stokes' law. Marine snow aggregates typically consist of inorganic particles, detrital organic particles, biological particles (e.g. mucus-feeding webs, molts, fecal pellets) as well as microorganisms [Alldredge and Silver, 1988]. Marine snow aggregates are fragile and disintegrate to primary particles when sampled by traditional means (nets, pumps, water samplers). The existence of occasionally significant concentrations of marine snow has been acknowledged for some time [Suzuki and Kato, 1953, Alldredge and Silver, 1988] and it is now generally believed that the downward flux of particulate material in the ocean occurs mainly in the form of such aggregates [Fowler and Knauer, 1986].

Aggregate formation of suspended particles may occur by physical coagulation. This is a process by which suspended particles collide due to fluid shear or differential settling velocities and stick together upon collision [e.g., McCave, 1984; Jackson and Lochmann, 1993]. The coagulation rate depend on the collision rate between suspended particles and on their "stickiness" (= probability of adhesion upon collision). The collision rate of spherical particles, in turn, is proportional to the particle size cubed, to the particle concentration squared and to the turbulent shear rate (thus ignoring Brownian motion and differential settling as mechanisms of particles encounter; these are insignificant for particles > 1 μm and in shallow water, respectively [McCave, 1984; Kiørboe et al., 1994]).

Sticky particles suspended in a fluid shear field will coagulate; as particles combine the concentration of particles will decline, and the average particle size will increase. Classical coagulation theory predicts that the decrease in particle concentration and the increase in average particle size over time will both be approximately exponential [Kiørboe et al., 1990a]. Figure 4.5 provides a laboratory demonstration, that phytoplankters may be sticky and that aggregate formation proceeds as

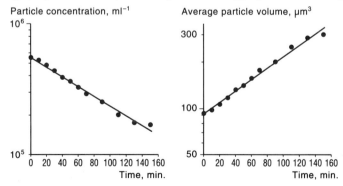

Figure 4.5. Cells of the diatom *Skeletonema costatum* will form aggregates by physical coagulation when suspended in a shear field. Due to coagulation the concentration of suspended particles will decline approximately exponential (a) and the average volume of particle aggregates will increase exponentially (b). In the present example cells were suspended in a laminar fluid shear field of 3 s^{-1} at a concentration of 50 ppm.

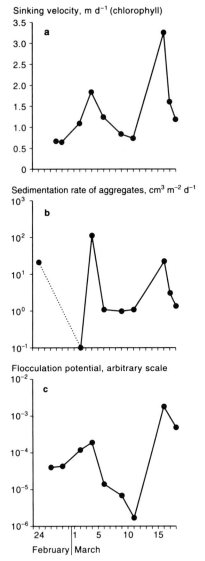

Sinking velocity, m d⁻¹ (chlorophyll)

a

Sedimentation rate of aggregates, cm³ m⁻² d⁻¹

b

Flocculation potential, arbitrary scale

c

February | March

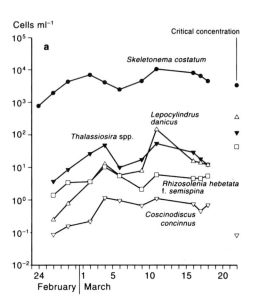

Cells ml⁻¹

a

Critical concentration

Skeletonema costatum

Lepocylindrus
danicus

Thalassiosira spp.

Rhizosolenia hebetata
f. semispina

Coscinodiscus
concinnus

February | March

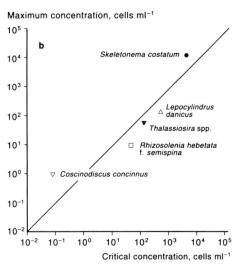

Maximum concentration, cells ml⁻¹

b

Skeletonema costatum

Lepocylindrus
danicus

Thalassiosira spp.

Rhizosolenia hebetata
f. semispina

Coscinodiscus concinnus

Critical concentration, cells ml⁻¹

Figure 4.6. Comparison of observed sinking velocity of suspended particles (a) and sedimentation rate of aggregates (b) with particle flocculation predicted from coagulation theory and measured particle stickiness and in situ turbulent dissipation rates (c) during a three-week period in a shallow Danish fjord (Isefjord). From Kiørboe et al. [1994].

Figure 4.7. Development of population sizes of five diatom species during a three-week period in a shallow Danish Fjord (Isefjord) (a). Populations of all species initially increase exponentially with similar population growth rates and, subsequently, stabilize at species-specific concentrations. These equilibrium concentrations are well predicted by coagulation theory (b). The line in (b) is $x = y$. From Kiørboe et al. [1994].

predicted by classical coagulation theory. Many marine suspended particles are sticky, including mineral particles [Edzwald et al., 1974; Leussen, 1988], TEP [Passow et al., 1994; Kiørboe and Hansen, 1993] and most diatoms [Kiørboe et al., 1990a; Kiørboe and Hansen; 1993, Logan et al., 1994; Drapeau et al., 1994]. Thus, coagulation is potentially important for the sedimentation of POM.

Figure 4.6 provides a field demonstration, that aggregate formation by coagulation can in fact account for observed patterns in sinking of particulate matter [Kiørboe et al., 1994]. In a shallow Danish fjord (Isefjord) a predictor of coagulation (based on measured concentrations and stickiness of suspended particles and on turbulent fluid shear rates calculated from wind velocities) mimicked very well the observed temporal pattern in particle sinking during a three-week period. This was further verified by the observed temporal pattern in occurrence and sedimentation rate of aggregated particles (Figure 4.6c). Thus, coagulation appears to be responsible for aggregation of suspended particles and an important determinant of sedimentation of particulate material in coastal waters.

Jackson [1990] combined classical coagulation theory with phytoplankton growth dynamics and showed that at a certain critical phytoplankton concentration, C_{cr}, cell growth and coagulation (and sedimentation) will balance, provided cell growth does not become nutrient-limited. An estimate of C_{cr} is given by:

$$C_{cr} = 0.384 \, \mu (\alpha \gamma d^3)^{-1} \tag{3}$$

where μ is the phytoplankton net growth rate in the absence of coagulation, α is the stickiness, d the cell diameter and γ the fluid shear rate. Thus, coagulation may also influence the population dynamics of the phytoplankton by putting an upper limit to cell concentrations. The development of the populations of five diatom species in the shallow Danish fjord (where nutrients did not become limiting for biomass accumulation) during a spring phytoplankton bloom is consistent with Jackson's critical concentration model: initially the populations grow exponentially and subsequently they stabilize at species-specific concentrations (Figure 4.7a). Also, the observed maximum cell concentrations of these five species were accurately predicted by the critical concentration model (Figure 4.7 b).

In the light of coagulation theory let us now reconsider the two above Kattegat examples where we found a discrepancy between observed and expected settling rates of particulate material:

The *Skeletonema costatum* bloom described by Olesen [1993] culminated with cell concentration of ca. 10^4 cells ml^{-1}, whereupon the cells aggregated and rapidly settled out; a significant fraction of the cells collected in sediment traps occurred as aggregates. How does this compare to the "critical concentration" model of Jackson? From the development of the phytoplankton population [Olesen, 1993] (Figure 4.3) we can estimate an initial net population growth rate (μ) of 0.06 d^{-1} (= $6.9 \cdot 10^{-7}$ s^{-1}). The wind velocity during the peak and decline of the bloom (February 20–28) averaged 6.5 m s^{-1} (observations from the Danish Meteorological Institute), and such a wind velocity gives rise to an average turbulent shear rate (γ) in the upper 10 m mixed layer of ca. 1 s^{-1} (calculated from the model presented by

MacKenzie and Leggett, 1993). With a length (d) of *S. costatum* cells of 7×10^{-4} cm and a cell stickiness (α) of 0.1 [Kiørboe et al., 1990a] we can calculate a critical concentration (C_{cr}) of 7.7×10^3 cells ml^{-1} by inserting these values in Eq. 3. This is close to the observed maximum cell concentration, 10^4 ml^{-1}, and coagulation theory, thus, appears capable of explaining the observed aggregation and sedimentation of the bloom.

In the second example considered above the higher than expected sinking velocity of suspended organic particles during summer in the Kattegat is also due to aggregation. Olesen and Lundsgaard [1994] found the majority of the particulate material collected in sediment traps to occur as large, amorphous aggregates. However, because aggregation by coagulation depends on the particle concentration squared, coagulation is not immediately consistent with the relatively low organic particle concentration encountered in the surface waters of the Kattegat during summer: 100–150 mg C m^{-3} (excl. zooplankton) or 1–1.5 ppm. The volume fraction of suspended particles may be much greater than measured microscopically or with an electronic particle counter because the particles of TEP (transparent exopolymeric particles) cannot be visualized in the microscope unless stained [Alldredge et al., 1993] and cannot be sensed by an electronic particle counter [Kiørboe and Hansen, 1993]. From the concentrations and average diameters of TEP in surface waters of the Santa Barbara Channel and Monterey Bay (California) reported by Alldredge et al. [1993] one can calculate that the volume fraction of suspended TEP here ranges between 0.1–400 ppm. Although we do not know to what extent these observations can be extrapolated to other coastal areas, including the Kattegat, they suggest that particle volume concentrations may be very much higher than measured by traditional techniques. And particle coagulation may be much more important than anticipated from the concentration and size distribution of "traditional" particles. This remains to be closer examined in the Kattegat and elsewhere, but may reconcile the observed aggregation and sedimentation of POM throughout the oligotrophic summer period in the Kattegat with classical coagulation theory.

4.9. Material Flux and Pelagic Food Web Structure in Coastal Waters

Because small organic particles do not sink significantly, and because POM does not accumulate in the water column, small organic particles are bound to be mineralized mainly in the pelagic zone. Small cells and particles are eaten by small heterotrophs, e.g. bacteria and picoalgae by heterotrophic nanoflagellates. These, in turn, are eaten by, for example, ciliates, that are eaten by copepods that are eventually eaten by consumers at higher trophic levels. Because there is a considerable respiratory energy loss at each trophic level, the majority of the primary production due to small cells is remineralized by microorganisms in this so-called microbial loop, and only little energy reaches higher trophic levels. Normally, small cells do not bloom, i.e., occur in very high biomass concentrations, because predator control is efficient. Exceptions are blooms of small toxic phytoplankters that may escape predation control by chemical defense, e.g. the bloom of the toxic hapto-phycean flagellate *Chrysochromulina polylepis* that occurred in the Kattegat during June 1988 [Nielsen et al., 1990]. However, small phytoplankton cells rarely occur in concentrations sufficient for coagulation and, hence, sedimentation to become important. Therefore the microbial loop is characteristic of regenerating systems where inorganic nutrients are efficiently recycled in the euphotic zone. There is a high degree of predator control in microbial communities; for example, population sizes of picoalgae and bacteria are controlled by their nanoflagellate grazers (cf. above and Andersen and Fenchel [1985]), and the ciliate community in the Kattegat is rarely if ever food-limited but rather limited efficiently by copepod predation [Nielsen and Kiørboe, 1991, 1994]. In the long run, however, the turnover rate in the regenerative microbial loop food webs appears to be governed by the light-limited growth rate of the phytoplankton.

Microbially dominated plankton communities are not always and exclusively based on regenerated production. In the Kattegat and the nearby Skagerrak, as well as elsewhere, the pycnocline is characterized by significant concentrations of auto-trophic nanoflagellates during summer stratification [Nielsen et al., 1990; Bjørnsen et al., 1993]. A significant fraction of the water column primary production is due to these subsurface populations, e.g. ca. 30% of the annual primary production in the Kattegat [Richardson and Christoffersen, 1991], and they apparently depend to a large extent on new, entrained nutrients [Bjørnsen et al., 1993; Olesen and Lundsgaard, 1994]. As neither POM nor remineralized nutrients accumulate here, there must be a downward flux of matter that equals the entrainment of new nutri-ent. Both in the Kattegat and Skagerrak such a downward flux has accordingly been observed [Bjørnsen et al., 1993; Olesen and Lundsgaard, 1994], and in the Kattegat the flux rate fits well with independent estimates of the rate at which new nutrients are entrained across the pycnocline [Olesen and Lundsgaard, 1994]. This sedimentation may be facilitated by packaging of small particles into larger fecal pellets. Sedimentation of small cells may also occur if the concentration of other sticky hydrosols, e.g. the particles of TEP, are sufficiently high for coagulation to occur at significant rates. In the Kattegat this is a likely mechanism by which new inorganic nutrients continuously entrained from below the euphotic zone or de-posited from the atmosphere again leave the euphotic zone, because sedimented material is characterized by aggregated particles (cf. above).

4.10. The Classical Food Chain: Grazing versus Sedimentation

Large phytoplankton cells bloom in spring and subsequent to episodic nutrient en-richments of the euphotic zone because their predators are insufficient in control-ling their population sizes. The primary production due to large phytoplankton cells fuel the so-called classical food chain: some fraction of the phytoplankton is consumed directly by mesozooplankters and thus made available to consumers at higher trophic levels. In the Kattegat and in the Skagerrak the production of cope-pods depends directly on the concentration and production of microalgae [Kiørboe

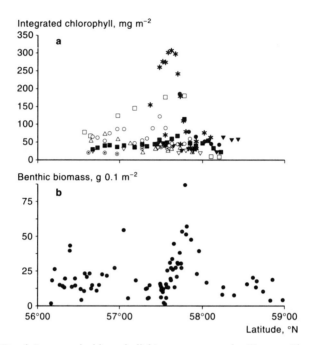

Figure 4.8. Depth-integrated chlorophyll biomass across the Kattegat/Skagerrak front as measured during seven cruises (a) and benthic invertebrate biomass (excluding large species) across the front (b). The position of the front is dynamic and varies between approximately 57°15 and 57°45 N. On average both phytoplankton and benthic biomasses are significantly elevated in the region covered by the excursions of the front. Phytoplankton data from Heilmann et al. [1994] and benthic invertebrate data from Josefson and Jensen (pers. comm.).

et al., 1990b; Kiørboe and Nielsen, 1994]. However, because large phytoplankton cells bloom exactly because predation pressure is low, and because the biomass concentration is high during blooms, the majority of the cells are likely to coagulate and subsequently settle out of the water column.

The spring phytoplankton bloom in temperate waters is dominated by large diatoms, and is an example of such an episodic bloom. In the Kattegat, for example, the mesozooplankton only consume < 1–5% [Nicolajsen et al., 1983] or ca. 12% [Kiørboe and Nielsen, 1994] of the spring bloom primary production. Grazing by heterotrophic dinoflagellates, the only other significant group of grazers that consume large phytoplankters [Hansen, 1991], may add to this. However, at least in some years, the majority of the spring bloom sediments to the seafloor in the Kattegat [e.g., Christensen and Kanneworff, 1985; Olesen 1993] as elsewhere (e.g. Kiel Bight [Smetacek, 1980; Peinert et al., 1982]).

Storm-induced entrainment of new nutrients to the surface layer and subsequent blooming of large phytoplankters is another example of episodic microalgal blooms. Kiørboe and Nielsen [1990] and Nielsen and Kiørboe [1991] described such a wind-generated bloom in the Kattegat. Although copepod productivity increased in response to the elevated food availability, the majority of the episodic, new production sedimented out of the water column. Other examples of short-lived wind-generated microalgal blooms and their effect on sedimentation and food web structure have been provided by e.g. Mullin et al. [1985], Cowles et al. [1987], Hitchcock et al. [1987], and Nair et al. [1989].

Fronts constitute the final example of localized and episodic availability of new nutrients [e.g., Loder and Platt, 1985] to be considered here. Fronts are frequently characterized by elevated concentrations of microalgae [Le Fèvre, 1986] and of high zooplankton productivity [Kiørboe et al., 1988], and these features also characterize the fronts between different water masses in the Kattegat/Skagerrak area [Richardson, 1985; Kiørboe et al., 1990] (see also Figure 4.8a). Elevated sedimentation of POM is also expected at fronts, but is difficult to document due to the dynamic position of most fronts. The Kattegat/Skagerrak front, for example, makes significant short-term excursions (on the order of 50–75 km [Pedersen and Møller, 1981]), and there are no sediment trap data available from the frontal areas in Danish waters. However, the biomass and productivity of benthic fauna is significantly elevated in the area covered by the excursions of the Kattegat/Skagerrak front relative to the regions north and south hereof (Figure 4.8b), suggesting elevated sedimentation of POM in the frontal region [Josefson and Jensen, unpublished].

4.11. Hydrodynamic Control of Pelagic Food Web Structure

The flux of material and the structure of pelagic food webs are ultimately governed by hydrodynamic processes. It has been shown above that

1. the spatio-temporal variation in the availability of "new" nutrients is under hydrodynamic control;
2. the size distribution of the phytoplankton is determined by the spatio-temporal variation in the availability of inorganic nutrients (continuous or pulsed); and
3. the fate of the pelagic primary production, flux of material and, hence, the structure of the pelagic food web depend on the size composition of the phytoplankton.

Thus, microbial food webs are characteristic of stratified waters, where the supply of nutrients is relatively continuous, and classical grazing food chains develop in shallow turbulent environments and at oceanographic discontinuities in water column structure, where nutrient availability is pulsed or episodic. The microbial food web and the classical grazing food chain represent extremes in a continuum, and in the sea all intermediate types of pelagic food web structure exist.

The export of organic material from the euphotic zone is primarily in the form of sedimenting particulate material and is eventually limited by the import of new inorganic nutrients. In microbial pelagic communities, where suspended particle mass is relatively low and the suspended particles small, coagulation and subsequent sedimentation of particulate material is presumably facilitated by the particles of TEP. It is otherwise difficult to reconcile observed sedimentation rates of amorphous organic aggregates in such communities with classical coagulation theory. In episodic, nutrient-rich environments the vertical flux of particulate material is substantial. Large particles (diatoms) occur in high concentrations, and aggregated live phytoplankton cells account for the major portion of the sedimented material.

These relationships are particularly evident on oceanwide and seasonal scales. For example, in the permanently stratified ocean gyres and during summer stratification in temperate waters, pelagic food webs appear to be of the microbial type. In the partially stratified zones of the major upwelling regions and during beginning vernal stratification in temperate waters, on the other hand, classical-type grazing food chains dominate and substantial sedimentation of phytoplankton occurs [Legendre and Le Fèvre, 1989; Kiørboe, 1991, 1993]. The latter type of environment also supports production at higher trophic levels and significant fisheries [e.g., Ryther, 1969; Cushing, 1989].

The variation in pelagic food web structure in relation to hydrodynamic features appears, however, to recur on much smaller spatio-temporal scales. Microbial food webs in stratified water columns may be rapidly replaced by grazing food chains and heavy sedimentation where and when episodic or localized injections of inorganic nutrients to the euphotic zone occur. Likewise, a microbial community again takes over once the short-lived microalgal bloom has exhausted the inorganic nutrients. Episodic storms and localized tidal fronts constitute examples, which were considered above. A transect study in the Skagerrak provides another example of how spatial variation in pelagic food web structure depends in a predictable way on km-scale variation in the hydrodynamics [Kiørboe et al., 1990b]. In the partially mixed peripheral regions of the Skagerrak microphytoplankton occurs in high concentration, copepod productivity (indicative of classical grazing food chain) is high and microbial activity (bacterial growth rate) is low; in the central, strongly stratified part, on the contrary, phytoplankton biomass is low and dominated by small flagellates, copepod productivity is low and microbial activity is high (Figure 4.9). Thus, in this example pelagic food web structure changes from a classical to a microbial and back to a classical type along a transect less than 80 km long.

4.12. Conclusion

Coastal ecosystems are in general physically more variable, energetic and turbulent than the open ocean. This is because tides are important, currents impinge on coasts and deposited wind energy will dissipate in a relatively shallow water column. The spatio-temporal variability in the hydrodynamics gives rise to a highly variable structure of the pelagic food web. Furthermore, the hydrodynamic processes cause injections of inorganic nutrients into the euphotic zone on an intermittent basis and to lead significant organic production that is bound to be eventually exported from the euphotic zone. Export of organic matter is mainly in the form of sedimentation to the sea floor and aggregate formation by physical coagulation appears to be important in facilitating the POM fallout.

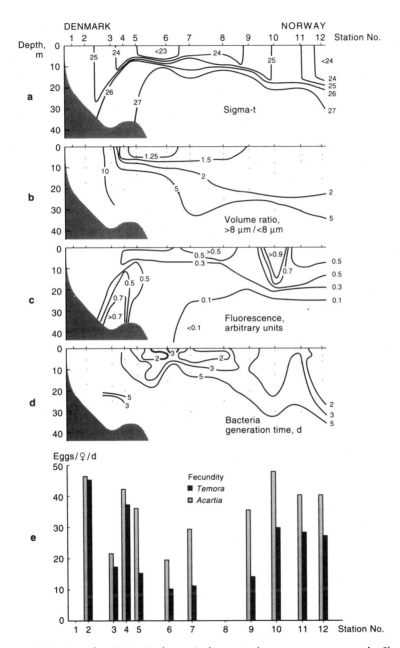

Figure 4.9. Horizontal variation in the vertical water column structure across the Skagerrak and associated variation in properties of the pelagic community in May. Water density as sigma-t units (a); volume ratio of large (> 8 μm) to small (< 8 μm) phytoplankters (b); chlorophyll biomass as fluorescence (c); bacteria generation time (= growth rate^{-1}) (d); and fecundities of the copepods *Acartia clausi* and *Temora longicornis*. From Kiørboe [1993].

The input of physical energy to coastal ecosystems will, on the one hand, act to homogenize the water column vertically. This is counteracted, however, by freshwater runoff from land, that will tend to vertically stabilize the water column. Coastal waters in general and estuaries – as the Kattegat – in particular are, therefore, typically strongly stratified due to salinity differences. The occurrence of a pycnocline and the different transport characteristics of solute and particulate material will vertically displace oxygen-consuming and -producing processes. Anthropogenic eutrophication may enhance the vertical flux of organic particles and, thus, potentially increase the oxygen deficit below the pycnocline with anoxia as one detrimental effect. However, the magnitude of the oxygen deficit is not related to the amount of introduced nutrients in a simple, linear way. As discussed in this chapter one reason for this is, that nutrients delivered intermittently are more likely to give rise to enhanced organic particle sedimentation than nutrients delivered continuously. In the latter case, remineralization of organic particles may occur within the euphotic zone, and other processes (e.g. denitrification) may remove "excess" inorganic nutrients from the water column.

Acknowledgments. Thanks are due to Paul Wassmann, Jens Heilmann, and Morten Søndergaard for critically reading an earlier version of the manuscript, and to Alf Josefsen for permission to use unpublished data. The manuscript was finalized in the summer of 1993.

References

Albert, M., and I. Valiela, Production of microbial organic aggregates from macrophyte-derived dissolved organic matter, *Limnol. Oceanogr.*, *39*, 37–50, 1994.

Alldredge, A. L., U. Passow, and B. Logan, The abundance and significance of a class of large, transparent organic particles in the ocean, *Deep-Sea Res.*, *40*, 1131–1140, 1993.

Alldredge, A. L., and M. W. Silver, Characteristics, dynamics and significance of marine snow, *Prog. Oceanogr.*, *20*, 41–82, 1988.

Andersen, P., and T. Fenchel, Bacterivory by microheterotrophic flagellates in seawater samples, *Limnol. Oceanogr.*, *30*, 198–202, 1985.

Baines, S. B., and M. L. Pace, The production of dissolved organic matter by phytoplankton and its importance to bacteria: Patterns across marine and freshwater systems, *Limnol. Oceanogr.*, *36*, 1078–1090, 1991.

Banse, K., Mass-scale rates of respiration and intrinsic growth in very small invertebrates. *Mar. Ecol. Prog. Ser.*, *9*, 281–297, 1982a.

Banse, K., Cell volumes, maximal growth rates of unicellular algae and ciliates, and the role of ciliates in the marine pelagial, *Limnol. Oceanogr.*, *27*, 1059–1079, 1982b.

Benner, R., S. D. Pakulski, M. McCarthy, J. I. Hedges, and P. G. Hatcher, Bulk chemical characterization of dissolved organic matter in the ocean, *Science*, *255*, 1561–1564, 1992.

Berg, H. C., and E. M. Purcell, Physics of chemoreception, *Biophys. J.*, *20*, 193–219, 1977.

Bjørnsen, P. K., Phytoplankton release of organic matter: Why do healthy cells do it?, *Limnol. Oceanogr. 33*, 151–154, 1988.

Bjørnsen, P. K., and J. Kuparinen, Heterotrophic dinoflagellate growth and herbivorous grazing in Southern Ocean microcosm experiments, *Mar. Biol. 109*, 397–405, 1991.

Bjørnsen, P. K., H. Kaas, H. Kaas, T. G. Nielsen, M. Olesen, and K. Richardson, Dynamics of a subsurface phytoplankton maximum in the Skagerrak, *Mar. Ecol. Prog. Ser.*, *95*, 279–294, 1993.

Blasco, D., T. T. Packard, and P. C. Garfidd, Size dependence of growth rate, respiratory electron transport system activity and chemical composition of marine diatoms in the laboratory. *J. Phycol.*, *18*, 58–63, 1982.

Børsheim, K. Y., Growth rate and mortality of bacteria in aquatic environments, Dissertation, The University of Trondheim, Norway, 63 pp., 1992.

Christensen, H., and E. Kanneworff, Sedimentating phytoplankton as major food source for suspension and deposit feeders in the Øresund, *Ophelia*, *24*, 223–244, 1985.

Cole, J. J., S. Findlay, and M. L. Pace, Bacterial production in fresh and saltwater ecosystems: a cross-system overview, *Mar. Ecol. Prog. Ser.*, *43*, 1–10, 1988.

Cowles, T., M. R. Roman, A. L. Ganzens, and N. J. Copley, Short-term changes in the biology of a warm-cove ring: zooplankton biomass and grazing, *Limnol. Oceanogr.*, *32*, 653–664, 1987.

Cushing, D. H., A difference in structure between ecosystems in strongly stratified waters and those that are only weakly stratified, *J. Plank. Res.*, *11*, 1–13, 1989.

Drapeau, D. T., H. G. Dam, and G. Grenier, An improved flocculator design for use in particle aggregation experiments, *Limnol. Oceanogr.*, *29*, 723–729, 1994.

Dugdale, R. C., and J. J. Goering, Uptake of new and regenerated forms of nitrogen in primary productivity, *Limnol. Oceanogr.*, *12*, 196–206, 1967.

Edzwald, J. K., J. B. Upchurch, and C. O. O'Melia, Coagulation in estuaries, *Envir. Sci. Technol.*, *8*, 58–63, 1974.

Fogg, G. E., The ecological significance of extracellular products of phytoplankton photosynthesis, *Botanica Marina*, *26*, 3–14, 1983.

Fowler, S. W., and G. A. Knauer, Role of large particles in the transport of elements and organic compounds through the oceanic water column, *Prog. Oceanogr.*, *16*, 147–194, 1986.

Geider, R. J., T. Platt, and J. A. Raven, Size dependence of growth and photosynthesis in diatoms, *Mar. Ecol. Prog. Ser.*, *30*, 93–104, 1986.

Goldman, J. C., On phytoplankton growth rates and particular C:N:P ratios at low light, *Limnol. Oceanogr.*, *31*, 1358–1363, 1987.

González, J. M., and C. A. Suttle, Grazing by marine nanoflagellates on viruses and virus-sized particles: ingestion and digestion, *Mar. Ecol. Prog. Ser.*, *94*, 1–10, 1993.

Hansen, P. J., Quantitative importance and trophic role of heterotrophic dinoflagellates in a coastal pelagic food web, *Mar. Ecol. Prog. Ser.*, *73*, 253–261, 1991.

Heilmann, J. P., K. Richardson, and G. Ærtebjerg, Annual distribution of phytoplankton in the Skagerrak/Kattegat frontal region, *Mar. Ecol. Prog. Ser.*, *112*, 213–223, 1994.

Hitchcock, G. L., C. Langdon, and T. J. Smayda, Short-term changes in the biology of a Gulf stream warm core ring: Phytoplankton biomass and productivity, *Limnol. Oceanogr.*, *32*, 919–928, 1987.

Jackson, G. A., A model of the formation of marine algal flocs by physical coagulation processes, *Deep-Sea Res.*, *37*, 1197–1211, 1990.

Jackson, G. A., and S. Lochmann, Modelling coagulation of algae in marine ecosystems, in *Environmental Particles*, edited by J. Buffle, vol. 2, pp. 373–399, 1993.

Jensen, L. M., and M. Søndergaard, Abiotic formation of particles from extracellular organic carbon released by phytoplankton, *Microbial Ecol.*, *8*, 47–54, 1982.

Jumars, P. A., D. L. Penry, J. A. Baross, M. J. Perry, and B. W. Frost, Closing the microbial loop: dissolved carbon pathway to heterotrophic bacteria from incomplete ingestion, digestion and absorption in animals, *Deep-Sea Res.*, *36*, 483–495, 1989.

Kahru, M., and A. Leeben, Size structure of planktonic particles in relation to hydrographic structure in the Skagerrak, *Mar. Ecol. Prog. Ser.*, *76*, 159–166, 1991.

Kepkay, P. E., S. E. H. Niven, and T. G. Milligan, Low molecular weight and colloidal DOC production during a phytoplankton bloom, *Mar. Ecol. Prog. Ser.*, *100*, 233–244, 1993.

Kiørboe, T., Pelagic fisheries and spatio-temporal variability in zooplankton productivity, *Bull. Plank. Soc. Japan*, Spec. vol., 229–249, 1991.

Kiørboe, T., Turbulence, phytoplankton cell size, and the structure of pelagic food webs, *Adv. Mar. Biol.*, 29, 1–72, 1993.

Kiørboe, T., and J. L. S. Hansen, Phytoplankton aggregate formation: observations of patterns and mechanisms of cell sticking and the significance of exopolymeric material, *J. Plank. Res.*, 15, 993–1018, 1983.

Kiørboe, T., and T. G. Nielsen, Effects of wind stress on vertical water column structure, phytoplankton growth, and productivity of planktonic copepods, in *Trophic Relationships in the Marine Environment*, edited by M. Barnes, and R. N. Gibson, pp. 28–40, Aberdeen University Press, Aberdeen, 1990.

Kiørboe, T, and T. G. Nielsen, Regulation of zooplankton biomass and production in a temperate, coastal ecosystem. I. Copepods, *Limnol. Oceanogr.*, 39, 493–507, 1994.

Kiørboe, T., P. Munk, K. Richardson, V. Christensen, and H. Paulsen, Plankton dynamics and herring larval growth, drift and survival in a frontal area, *Mar. Ecol. Prog. Ser.*, 44, 205–219, 1988.

Kiørboe, T., K. P. Andersen, and H. G. Dam, Coagulation efficiency and aggregate formation in marine phytoplankton, *Mar. Biol.*, 107, 235–245, 1990a.

Kiørboe, T., H. Kaas, B. Kruse, F. Møhlenberg, P. Tiselius, and G. Ærtebjerg, The structure of the pelagic food web in relation to water column structure in the Skagerrak, *Mar. Ecol. Prog. Ser.*, 59, 19–32, 1990b.

Kiørboe, T., C. Lundsgaard, M. Olesen, and J. L. S. Hansen, Aggregation and sedimentation processes during a spring phytoplankton bloom: a field experiment to test coagulation theory, *J. Mar. Res.*, 52, 1–27, 1993.

Koike, I., S. Hara, T. Terauchi, and K. Kogure, Role of sub-micrometer particles in the ocean, *Nature*, 345, 242–244, 1990.

Le Fèvre, J., Aspects of the biology of frontal systems, *Adv. Mar. Biol.*, 23, 163–299, 1986.

Legendre, L., and J. Le Fèvre, Hydrodynamical singularities as controls of recycled versus export production in oceans, in *Productivity of the Ocean: Present and Past*, edited by W. H. Berger, V. Smetacek, and G. Weber, pp. 49–63, 1989.

Loder, W. J., and T. Platt, Physical controls on phytoplankton production at tidal fronts, in *Proc. 19th Europ. Mar. Biol. Symp.*, edited by P. E. Gibbs, pp. 3–19, Cambridge University Press, Cambridge, 1985.

Logan, B. E., and A. L. Alldredge, Potential for increased nutrient uptake by flocculating diatoms, *Mar. Biol.*, 101, 443–450, 1989.

Logan, B. E., U. Passow, and A. L. Alldredge, Variable retention of diatoms on screens during size separation, *Limnol. Oceanogr.* 390–395, 1994.

Longhurst, A. R., I. Koike, W. K. W. Li, J. Rodriguez, P. Dickie, P. Kepay, F. Partensky, B. Bautista, J. Ruiz, M. Wells, and D. F. Bird, Sub-micron particles in northwest Atlantic Shelf water, *Deep-Sea Res.*, 39, 1–7, 1992.

MacKenzie, B. R. and W. C. Leggett, Wind-based models for estimating the dissipation rates of turbulent energy in aquatic environments: empirical comparisons, *Mar. Ecol. Prog. Ser.*, 94, 207–216, 1993.

Marchant, H. J., and F. J. Scott, Uptake of sub-micrometer particles and dissolved organic material by Antarctic choanoflagellates, *Mar. Ecol. Prog. Ser.*, 92, 59–64, 1993.

McCave, I. N., Size spectra and aggregation of suspended particles in the deep ocean, *Deep-Sea Res.*, 31, 329–352, 1984.

Moral, F. M. M., and P. M. Gschwend, The role of colloids in the partitioning of solutes in natural waters, in *Aquatic Surface Chemistry*, edited by W. Stumm, pp. 405–422, John Wiley & Sons, New York, 1987.

Mullin, M. M., E. R. Brooks, F. M. H. Reid, J. Napp, and E. F. Stewart, Vertical structure of nearshore plankton off southern California: a storm and a larval fish food web, *Fish. Bull. U.S.*, 83, 151–170, 1985.

Munk, W. H., and G. A. Riley, Absorption of nutrients by aquatic plants, *J. Mar. Res.*, 11, 215–240, 1952.

Murray, A. G., and G. A. Jackson, Viral dynamics: a model of the effects of size, shape, motion and abundance of single-celled planktonic organisms and other particles, *Mar. Ecol. Prog. Ser.*, 89, 103–116, 1992.

Nair, R. R., V. Ittekkot, S. J. Manganini, V. Ramaswamy, B. Haake, E. T. Degens, B. N. Desai, and S. Honjo, Increased particle flux to the deep ocean related to monsoons, *Nature* (Lond.), 338, 749–751, 1989.

Nicolajsen, H., F. Møhlenberg, and T. Kiørboe, Algal grazing by the planktonic copepods *Centropages hamatus* and *Pseudocalanus* sp.: Diurnal and seasonal variation during the spring phytoplankton bloom in the Øresund, *Ophelia*, 22, 15–31, 1983.

Nielsen, T. G., and T. Kiørboe, Effects of a storm event on the structure of the pelagic food web with special emphasis on planktonic ciliates, *J. Plank. Res.*, 13, 35–51, 1991.

Nielsen, T. G., and T. Kiørboe, Regulation of zooplankton biomass and production in a temperate, coastal ecosystem. II. Ciliates, *Limnol. Oceanogr.*, 39, 508–519, 1994.

Nielsen, S. L., and K. Sand-Jensen, Allometric scaling of maximal photosynthetic growth rate to surface/volume ratio, *Limnol. Oceanogr.*, 35, 177–181, 1990.

Nielsen, T. G., T. Kiørboe, and P. K. Bjørnsen, Effect of a *Chrysochromulina polylepis* sub-surface bloom on the plankton community, *Mar. Ecol. Prog. Ser.*, 62, 21–35, 1990.

Olesen, M., The fate of an early diatom spring bloom in the Kattegat, *Ophelia*, 37, 51–66, 1993.

Olesen, M. and C. Lundsgaard, Seasonal sedimentation of autochthonous material from the euphotic zone of a coastal ecosystem, *Estuar. Coast. Mar. Sci.*, 1995.

Passow, U., A. L. Alldredge, and B. E. Logan, The role of particulate carbohydrate exudates in the flocculation of diatom bloom, *Deep-Sea Res.*, 41, 335–357, 1994.

Peinert, R., A. Saure, P. Stegman, C. Stienen H. Haardt, and V. Smetacek, Dynamics of primary production in coastal ecosystems, *Neth. J. Sea Res.*, 16, 276–289, 1982.

Richardson, K., Plankton distribution and activity in the North Sea/Skagerrak-Kattegat frontal area in April 1984, *Mar. Ecol. Prog. Ser.*, 26, 233–244, 1985.

Richardson, K., and A. Christoffersen, Seasonal distribution and production of phytoplankton in the southern Kattegat, *Mar. Ecol. Prog. Ser.*, 78, 217–227, 1991.

Riebesell, U., Comparison of sinking and sedimentation rate measurements in a diatom winter/spring bloom, *Mar. Ecol. Prog. Ser.*, 54, 109–119, 1989.

Ryther, J. H., Photosynthesis and fish production in the sea, *Science*, 166, 72–76, 1969.

Sheldon, R. W., A. Prakash, and W. H. Sutcliffe jr., The size distribution of particles in the ocean, *Limnol. Oceanogr.*, 17, 327–340, 1972.

Shimeta, J., Diffusional encounter of submicrometer particles and small cells by suspension feeders, *Limnol. Oceanogr.*, 38, 456–465, 1993.

Smayda, T. J. and B. J. Boleyn, Experimental observations on the flotation of marine diatoms. II. *Skeletonema costatum* and *Rhizosolenia setigera*, *Limnol. Oceanogr.*, 11, 18–34, 1966.

Smetacek, V., Annual cycle of sedimentation in relation to plankton ecology in western Kiel Bight, *Ophelia*, Suppl. 1, 65–76, 1980.

Smetacek, V., Role of sinking in diatom life-history: ecological, evolutionary and geological significance, *Mar. Biol.*, 84, 239–251, 1985.

Smith, S. V., and J. T. Hollibaugh, Coastal metabolism and the oceanic organic carbon balance, *Rev. Geophys.*, 31, 75–89, 1993.

Sommer, U., Some size relationships in phytoflaggelate motility, *Hydrobiologia*, *161*, 125–131, 1988.

Søndergaard, M., and M. Middelboe, A cross-system analysis of labile dissolved organic carbon, *Mar. Ecol. Prog. Ser.*, 283–294, 1995.

Stachowitsch, M., N. Fanuko, and M. Richter, Mucus aggregates in the Adriatic Sea: An overview of stages and occurrences, *Mar. Ecol.*, *11*, 327–350, 1990.

Strayer, D., On the limits to secondary production, *Limnol. Oceanogr.*, *33*, 1217–1220, 1988.

Sugimura, Y., and Y. Suzuki, A high temperature catalytic oxidation method of non-volatile dissolved organic carbon in seawater by direct injection of liquid sample, *Mar. Chem.*, *24*, 105–131, 1988.

Suzuki, N., and K. Kato, Studies on suspended materials, Marine snow in the sea. I. Sources of marine snow, *Bull. of the Faculty of Fisheries of Hokkaido University*, *4*, 132–135, 1953.

Thomsen, H. A. (Ed.), *Plankton in the Inner Danish waters* (in Danish), *Havforskning fra Miljøstyrelsen*, *11*, 331 pp., Danish Environmental Protection Agency, Copenhagen, 1992.

Tranvik, C. J., E. B. Sherr, and B. F. Sherr, Uptake and utilization of "colloidal DOM" by heterotrophic flagellates in seawater, *Mar. Ecol. Prog. Ser.*, *92*, 301–309, 1993.

van Leussen, W., Aggregation of particles, settling velocities of mud flocs, A review, in *Physical Processes in Estuaries*, edited by J. Dronkers and W. van Leussen, pp. 347–403, Springer, Berlin, 1988.

von Bodungen, B., K. V. Brockel, V. Smetacek, and B. Zeitzschel, Growth and sedimentation of the phytoplankton spring bloom in the Bornholm Sea (Baltic Sea), *Kieler Meeresforsch.*, *5*, 49–60, 1981.

Wells, M. L., and E. D. Goldberg, Occurrence of small colloids in sea water, *Nature*, *353*, 342–344, 1991.

Williams, P. J., LeB., Incorporation of microheterotrophic processes into the classical paradigm of the planktonic food web, *Kieler Meeresforsch*, Sonderheft 5, 1–28, 1981.

5

Carbon Flow in the Water Column
Case Study: The Southern Kattegat

Katherine Richardson

5.1. Introduction

The Kattegat is among the best studied marine regions in the world and certainly one for which some of the longest time series for biological data are available. Nevertheless, the increase in intensity and geographic distribution of anoxia and hypoxia observed in this sea during the 1980s came as a surprise to scientists as well as to the public at large. Records from the early part of this century suggest that certain areas in the Kattegat (especially in the Belt Seas) have long been plagued by periodic occurrences of hypoxia. However, previously these incidents were apparently limited to deeper "holes" in the sea bottom where bottom water exchange can be restricted – especially during periods of calm weather (typically in late summer).

Since the early 1980s, widespread hypoxia (defined here as oxygen concentrations of < 3 mg O_2 l^{-1}) has become an annual feature in bottom waters of the Kattegat. The magnitude of the problem varies, however, from year to year, presumably as a function of weather (mixing) conditions and nutrient input. The most severe conditions observed to date were recorded in 1988 where the hypoxia extended essentially over the entire Kattegat south of about 57°N latitude (see chapter 1; Figure 1.6).

It has generally been assumed [Ærtebjerg, 1987; Rosenberg et al., 1990; Kronvang et al., 1993] that the cause of the observed increase in oxygen depletion of bottom waters is an increase in sedimentation of organic material from the water column to the bottom which gives rise to an increase in biological oxygen demand (BOD) in near-bottom waters. This presumed increase in sedimentation of organic material should be a consequence of an increase in phytoplankton primary production resulting from a documented increase in eutrophication of this area (see chapters 3 and 12).

Eutrophication in Coastal Marine Ecosystems
Coastal and Estuarine Studies, Volume 52, Pages 95–114
Copyright 1996 by the American Geophysical Union

Primary production in this and most other estuarine marine areas has been shown to respond directly to or be correlated with increases in nitrogen input [Granéli, 1987; Nixon, 1992]. Thus, it is increases in nitrogen loading that are frequently cited as being responsible for hypoxia in the Kattegat. Very little was actually known, however, about carbon flow within the water column here or about any changes that may have occurred in the pattern of this flow as a result of eutrophication prior to the execution of the Danish Marine Research Program during the period 1988–1993.

A major goal of this program was to increase understanding of the interaction between nutrient input and oxygen depletion in Kattegat bottom waters in order to determine whether a reduction in nutrient loading can be expected to reduce the intensity and/or frequency of hypoxic events. In order to achieve this goal, it was necessary to improve the existing understanding of energy flow within the pelagic community. Therefore, a number of studies were initiated which contributed to the overall goals of

1. Describing the annual cycling of carbon, nitrogen and oxygen within the water column
2. Identifying the potential influence of hydrographic features and in particular, wind-driven nutrient upwelling on patterns of energy flow and
3. Identifying any long-term changes that may have occurred in the production of organic material in the Kattegat.

5.2. Description of the Studies Carried Out

One central study was carried out at a fixed position in the southern Kattegat (56°11' N; 12°04' E) between 1988 and 1990. The sampling strategy was designed to provide a description of carbon flow within this system for the calendar year 1989. A buoy equipped with automated temperature, salinity and current sensors and a fluorometer for determining chlorophyll-a fluorescence was deployed at the station. Sedimentation was estimated using sediment traps throughout the water column. These, however, were unfortunately only in position during the period March to October 1989 [Olesen, 1993; Olesen and Lundsgaard, 1994; Olesen, 1994].

In addition to the automatic registration of data carried out at this position, the station was visited at approximately fortnightly intervals. On these visits, standard hydrochemical determinations were carried out. Samples were also taken for the determination of biomass and/or speciation of bacteria [Bjørnsen et al., 1992], phytoplankton, heterotrophic nanoflagellates [Thomsen et al., 1992] microzooplankton [Hansen, 1991; Nielsen and Kiørboe, 1994] and mesozooplankton [Kiørboe and Nielsen, 1994]. Primary [Richardson and Christoffersen, 1991] and secondary [Bjørnsen et al, 1992: bacteria; Kiørboe and Nielsen, 1994: mesozooplankton] production measurements were also carried out on most visits.

Short-term fluctuations in carbon flow were examined during three "intensive" study periods in which the buoy position was visited every day (weather permitting) during a period of approximately three weeks. These took place in May–June, 1988, August–September, 1989 and October–November, 1989. During a 10-day period in April 1989 and the August–September 1989 intensive sampling period, measurements of nitrate uptake using standard ^{15}N techniques [Kaas et al, 1992] were carried out.

5.3. Carbon/Nitrogen Flow: Annual Cycle

An estimate for phytoplankton primary production of 290 g C m^{-2} yr^{-1} was made using the ^{14}C incorporation method at the fixed position in the Kattegat in 1989 [Richardson and Christoffersen, 1991]. The exact relationship between ^{14}C incorporation and net and gross photosynthesis is unclear [Peterson, 1980] but in another study carried out in this research program, Jespersen et al. [1994] found using several laboratory cultures of phytoplankton that the results from the ^{14}C method approximated net photosynthesis under most of the conditions tested.

Further studies are necessary in order to determine the relationship between ^{14}C incorporation and net/gross photosynthesis under natural conditions. However, even if the ^{14}C method of estimating primary production approximates net photosynthesis as suggested by Jespersen et al., the value of 290 g C m^{-2} yr^{-1} for annual primary production at the fixed station in the Kattegat presented by Richardson and Christoffersen [1991] must be corrected for phytoplankton's dark respiration before an estimate of the carbon available for the pelagic food chain can be made.

Net diel photosynthesis (i.e. that occurring over a 24-hour period) cannot be directly measured. From culture studies, however, it is known that dark respiration is often on the order of 5% of the maximum rate of photosynthesis. The light conditions over much of the water column do not, of course, support maximum photosynthesis. Given these considerations, a respiration loss due to dark respiration on the order of 15% of the total primary production would appear reasonable. Using this correction, phytoplankton respiration would account for a consumption of about 45 g C m^{-2} yr^{-1} at the buoy station. A reasonable estimate of net primary production would seem, then, to be on the order of 245 g C m^{-2} yr^{-1}.

Sedimentation of organic material at the buoy station was estimated to be 63 g C m^{-2} during the period March to October, 1989 [Olesen and Lundsgaard, 1994]. It was not possible to maintain the sediment traps in position during the winter period. However, sedimentation during the winter months is expected to be small. In addition, the sedimentation values during the period March to October agree well with an earlier estimate of sedimentation in the Kattegat (66 g m^{-2} yr^{-1}) made on the basis of benthic metabolism at a number of different stations [Jørgensen and Revsbech, 1989]. Thus, the value of 63 g C m^{-2} is treated here as an estimate of annual sedimentation occurring at the buoy station. Some of this sedimented material derives directly from primary production while some will, of course, originate from heterotrophic processes.

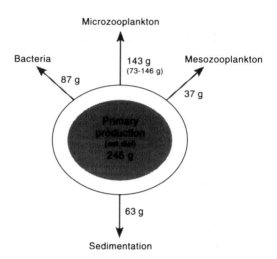

Figure 5.1. Flow of carbon in the pelagic food web at a fixed position in the Kattegat (56°11'N; 12°04'E). All units are in g yr^{-1}. Note that the sum of the values outside of the ring is greater than the value for net primary production. This is because some of the carbon being consumed by the various compartments or sedimenting is the product of heterotrophic processes. Thus, carbon fixed via primary production can appear more than once in the description of carbon flow. See text for the sources of the values cited.

Kiørboe (see chapter 4, this volume) has emphasized that a quantitative description of the flow of carbon (energy) within the pelagic community would require estimates of respiratory loss of carbon at each trophic level because of the potential confusion resulting from loss and reuse of carbon at all trophic levels. Unfortunately, such estimates do not exist. Nevertheless, a semiquantitative description of the flow of carbon from primary producers and into the pelagic food chain in the Kattegat can be obtained by examining the ingestion by the various pelagic compartments (Figure 5.1).

Bjørnsen et al. [1992] estimated bacterial carbon consumption in the water column to be on the order of 30% of primary production (ca. 87 g C m^{-2} yr^{-1}) at the buoy position in 1989. Kiørboe and Nielsen [1994] and Nielsen and Kiørboe [1994] have estimated grazing by mesozooplankton and microzooplankton (ciliates) at this station to be 37 and 143 g C m^{-2} yr^{-1}, respectively.

Other workers [Hansen and Nielsen, 1992] estimated that microzooplankton grazing consumed 10–20% of the phytoplankton biomass on a daily basis. Phytoplankton biomass at this station in 1989 converted from chlorophyll to carbon units has been presented by Kiørboe (chapter 4, this volume). With the exception of the spring bloom period when it reaches a value of nearly 25 g C m^{-2}, phytoplankton biomass is on the order of 2 g m^{-2} during the productive period of the year and slightly less during the winter. Applying a value of 2 g C m^{-2} to Hansen and Nielsen's estimate of microzooplankton consumption, yields a value for microzooplankton grazing of 73–146 g C m^{-2} yr^{-1}.

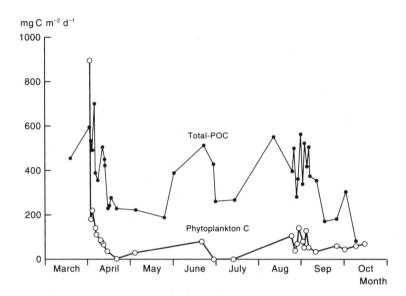

Figure 5.2. Sedimentation of total particulate organic carbon (POC) and phytoplankton carbon from the surface layer through the pycnocline into the bottom water at the buoy station in the Kattegat: March–October, 1989. After Olesen and Lundsgaard [1992].

As pointed out earlier, carbon ingestion by these various compartments of the ecosystem need not be mutually exclusive. Bacteria do not, for example, draw all of their carbon directly from primary producers and it is not only phytoplankton particles that are ingested by micro- and mesozooplankton. Nevertheless, this analysis suggests that most of the phytoplankton primary production occurring at the Kattegat station is processed in the water column. Sedimenting material, then, will largely be the products of heterotrophic processes. This argument is supported by a comparison of phytoplankton carbon and total POC sedimenting through the pycnocline from the surface to bottom layer (Figure 5.2) which indicates that POC sinking out is only dominated by phytoplankton C during the spring bloom. During the rest of the productive period, phytoplankton C (as determined from the chlorophyll content in the sedimenting material) represents < 30% of POC [Olesen and Lundsgaard, 1994].

5.4. Nitrogen Input and Primary Production

Phytoplankton production in marine waters has often been shown to be limited by nitrogen availability. In addition, total primary production is comprised of a "new" and a "regenerated" component depending upon the ionic nitrogen source used to support the production [Dugdale and Goering, 1967]. Only "new" production (based on nitrate-N) can lead to a net increase in organic material in the system. Thus, it is new production that is most relevant to consider in terms of eutrophication analyses. A quantitative distinction between new and regenerated production in these coastal waters is not possible. Nevertheless, an estimate of the magnitude of the net increase in the production of organic material in the Kattegat can be made by considering the total nitrogen availability in the Kattegat.

The annual net loading of nitrogen to the Kattegat via land runoff, atmospheric deposition and advection from surrounding seas has been estimated to be on the order of 1.5×10^5 t [Planlægningsrådet, 1987; Hansen et al., 1994; and chapters 3 and 12, this volume]. In addition to the nitrogen entering the Kattegat from external sources, there is a pool of nitrogen already present in Kattegat waters. From inorganic nitrogen concentrations in winter (when we assume that the majority of the nitrogen present in the pelagic system is in the inorganic form), this pool size can be estimated to be on the order of 1×10^5 t N [Richardson and Christoffersen, 1991]. Dividing by the area (31,000 km^2) applying the Redfield [1958] ratio (C:N molar ratio = 5.7) to the combined allochthonous and autochthonous nitrogen sources and assuming that all of these nitrogen atoms are used just once during the annual production cycle yields the theoretical new production in this area (48 g C m^{-2} yr^{-1}).

Comparison of this value for new production and total annual production (290 g C m^{-2} yr^{-1} before correction for phytoplankton's dark respiration) leads to the conclusion that most of the nitrogen within the system is being remineralized and reused during an annual production cycle. The measured sedimentation rate (63 g C m^{-2}) suggests that this reuse leads not only to regenerated production [sensu Dugdale and Goering, 1967] but also to some new production (i.e. leads to a net increase in organic material). That the theoretically derived estimate of new production may be an underestimate of actual new production may be at least partly explained by the fact that, in the above analysis, the Kattegat is treated as a single system and nitrogen input to the system as a whole is considered. In reality, however, phytoplankton activity in the Kattegat is controlled by nutrient exchanges occurring at a much more local scale than that of the entire Kattegat (Figure 5.3).

This region is strongly stratified during the most productive part of the year. The surface layer corresponds more or less to the euphotic zone and the phytoplankton activity in the surface layer reduces the availability of nutrients in this layer so that nutrients become limiting for primary production. In their sedimentation study, Olesen and Lundsgaard [1994] noted that nitrogen is apparently remineralized faster in the water column than carbon at the buoy station as evidenced by the fact that sedimenting organic material exhibited a higher C:N ratio (8.4 molar ratio) than that observed for suspended material in the photic zone (6.6). They suggest that at least some of this water column remineralization is occurring above the pycnocline and this decoupling of N and C would, in theory, mean that an individual nitrogen molecule may support "extra" phytoplankton production which can be considered as new production with respect to carbon but as regenerated with respect to nitrogen.

In addition to remineralization of nitrogen in the upper part of the water column, the transfer of nitrate from bottom to surface waters (and the atmospheric deposition of nitrogen to surface waters) will also lead to new production. Wind-generated

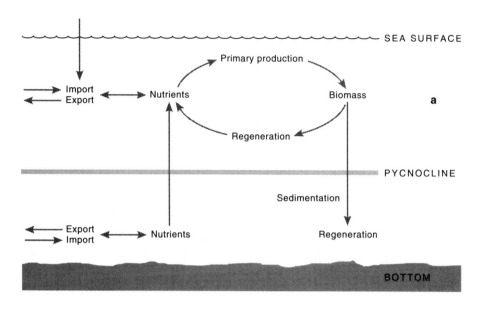

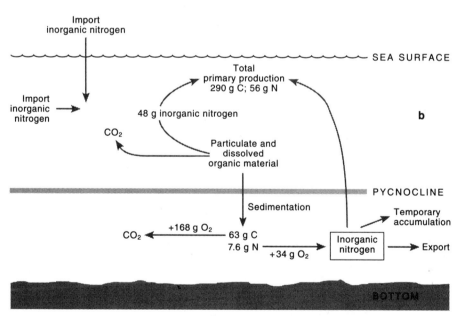

Figure 5.3. a: Schematic diagram identifying sources of nutrients (including location of nutrient regeneration in the water column) which control phytoplankton activity in the Kattegat. b: Biologically mediated carbon and nitrogen flow in the water column at the fixed
position in the Kattegat during 1989. The amount of oxygen required to remineralize the
sedimented material is also given in the figure. Sources for the values presented are given in
the text.

mixing of regenerated nitrate from the bottom to the surface layer is likely to contribute to new production throughout the summer season. Hansen et al. [1990] have estimated that on the order of 1.1×10^5 t N were transported annually from bottom waters through the pycnocline and into surface waters during the summer period during the 1980s. This value would correspond to a net organic production of 29 g C m^{-2} yr^{-1}. In addition, it must be expected that the primary production occurring during the winter, at the time of the spring bloom and in the bottom waters (when the photic zone extends to below the pycnocline) will largely be supported by nitrate and, thus, be characterized by a relatively large proportion of "new" production.

5.5. Oxygen Budget: Bottom Waters

Data are available for the period of most active phytoplankton production (March–October) in 1989 that allow examination of the oxygen budget in bottom waters at the buoy station. Olesen and Lundsgaard [1994] estimated that the organic carbon sedimentation of 63 g m^{-2} here during this period was accompanied by the sedimentation of 7.4 g of nitrogen. This sedimented carbon would require 168 g O$_2$ during its respiration to CO$_2$ and the nitrogen would require 34 g O$_2$ to be oxidized to nitrate. Thus, a total of 202 g O$_2$ m^{-2} would be required for the remineralization of the sedimented material (Figure 5.3b).

A reduction in oxygen concentration of only 61 g O$_2$ m^{-2} in bottom water (i.e. below the pycnocline) was recorded during the same period [Olesen and Lundsgaard, 1994]. However, there is an input of oxygen to bottom waters during the summer months via lateral advection from more oxygen-rich areas (e.g. Skagerrak) and by mixing through the pycnocline which will, to some extent, mitigate the oxygen loss due to remineralization processes. Models [Hansen et al., 1990; Hansen et al., 1994; and chapter 11, this volume] describing water flow and mixing processes in this region suggest that an oxygen usage below the pycnocline of approximately 190 g m^{-2} in remineralization combined with the oxygen input via hydrographic processes would yield a drop in oxygen concentration in bottom waters that agrees well with the observed. This good agreement between measured changes in oxygen concentration in the bottom waters and those predicted from carbon sedimentation rates and oxygen advection supports the assumed link between oxygen depletion events and primary production (sedimentation of organic carbon).

5.6. Seasonal Cycle in Sedimentation

It has previously been assumed that the major sedimentation of carbon from the water column to the bottom occurs in connection with the spring bloom. However, one of the results that emerged from this study was that the sedimentation of particulate organic carbon out of the water column was variable but remained high from the time of the spring bloom until late summer (Figure 5.2). During the summer period when phytoplankton carbon was a minor contributor to the total POC, microscopic examination of sedimented material indicated that it was largely comprised of "aggregates" or marine snow of unidentified origin [Olesen and Lundsgaard, 1994].

5.7. Subsurface Phytoplankton Peak

Another result arising from the study was the fact that a subsurface phytoplankton peak dominates primary production in the Kattegat from the period immediately following the spring bloom until the autumnal breakdown of stratification in September–October [Richardson and Christoffersen, 1991; Heilmann et al., 1994].

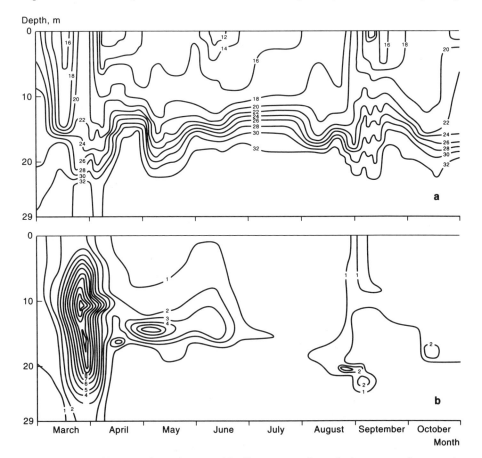

Figure 5.4. Distribution of a) salinity and b) fluorescence through the water column at the buoy station in the Kattegat from March to October, 1989. After Olesen and Lundsgaard [1992].

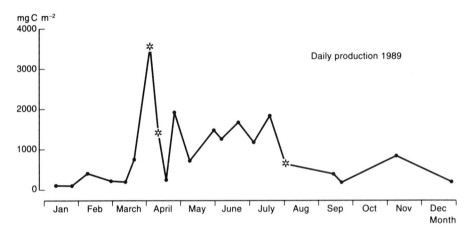

Figure 5.5. Daily primary production at the buoy station in the Kattegat during 1989. Stars represent weeks in which more than one sample was taken. From Richardson and Christoffersen [1991].

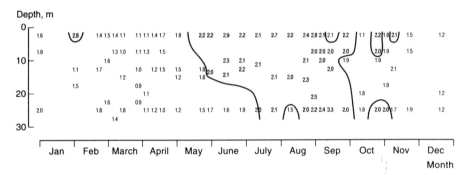

Figure 5.6. Distribution of the 480/665 absorption ratio through the water column at the buoy station in 1989. Values of over 2.0 (shaded) indicate a phytoplankton population that is nutrient-depleted [see Heath et al., 1990]. From Richardson and Christoffersen [1991].

The occurrence of subsurface chlorophyll peaks in this area has been acknowledged for over fifty years [Pettersson, 1934]. However, the studies described here represent the first attempt to quantify the importance of these peaks for the total primary production taking place in the Kattegat.

Figure 5.4 shows the vertical distribution of chlorophyll throughout the water column at the buoy station from March to October, 1989. The spring bloom is clearly the most significant feature in the surface waters at this station. However, a strong subsurface chlorophyll peak was present from the end of March until the middle of July. Although chlorophyll concentrations were low throughout the water column during the period between the middle of July and the middle end of August, Richardson and Christoffersen [1991] have reported that a subsurface chlorophyll peak was also present during this period.

The effect of the subsurface phytoplankton peak from the period following the spring bloom until the middle of July can be seen in the annual distribution of primary production throughout the year in 1989 (Figure 5.5). Although the highest daily rates recorded during the year were during the very short spring bloom, substantial rates were also recorded in the period following the spring bloom until the middle of July as a result of the activity of the subsurface phytoplankton peak. Richardson and Christoffersen [1991] have estimated that 19% of the annual primary production at the buoy station occurred during the spring bloom while on the order of 33% occurred in the subsurface peak in the months following the spring bloom.

The ratio of carotenoid to chlorophyll pigments (as indicated by the absorption in acetone extracts at 480 and 665 nm) can be used as a qualitative indicator of the nutrient status of a phytoplankton population [Heath et al., 1990]. The distribution of this ratio throughout the water column (Figure 5.6) shows the presence of high values (indicating nutrient depletion) in surface waters and low values in bottom waters from the period following the spring bloom until about mid-July. After this time and until the autumnal breakdown of stratification, high ratios are recorded throughout the water column, suggesting the presence of nutrient depleted phytoplankton throughout the water column.

The light attenuation characteristics of the water column show that the lower boundary of the euphotic zone (1% light penetration level) in this region during July and August lies between 15 and 18 m [Richardson and Christoffersen, 1991]. The annual distribution of nitrate throughout the water column at a nearby station in the Kattegat (Figure 5.7) indicates that the nutricline deepens throughout the summer. Measurements of nitrate concentrations carried out in 1989 at a station

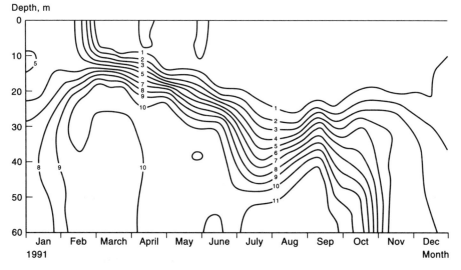

Figure 5.7. Annual distribution of nitrate through the water column during 1991 at a fixed station (56° 40'N; 12° 07'E) in the Kattegat. From Ærtebjerg [1992].

TABLE 5.1. Nitrate concentrations (μmol l^{-1}) at 15 and 20 m from May to Sept. 1989 near the buoy station (see text). Data supplied by the Danish National Environmental Research Agency.

	29 May	11 July	14 Aug	18 Sep
15 m	0.2	0.5	0	0.3
20 m	12.4	11.5	0.7	1.3

TABLE 5.2. Nitrate concentrations (μmol l^{-1}) at 15 and 20 m during July and August at a station in the southern Kattegat during the 1980s (see text for details). Data supplied by the Danish National Environmental Research Agency.

Year	Depth	July	August
1981	15 m	0.5	0
	20 m	0.2	0.4
1982	15 m	0	0
	20 m	0.7	0
1983	15 m	0	0.1
	20 m	10.6	2.1
1984	15 m	0.2	0
	20 m	0.1	0
1985	15 m	0.1	0
	20 m	0.1	0.1
1986	15 m	0.7	0
	20 m	1.9	0.5
1987	15 m	0.1	0.5
	20 m	1.9	0.5
1988	15 m	No data	0
	20 m	No data	0
1989	15 m	0.5	0
	20 m	8.0	0

near the buoy station (56°10' N, 11°48' E) show a very substantial drop in the concentration of nitrate at 20 m between mid-July and mid-August (Table 5.1). A longer time series of nitrate concentrations at 15 and 20 m during the months of July and August at another station in the southern Kattegat (56°40'N; 12°07'E: Table 5.2) suggest that this seasonal pattern in the distribution of nitrate throughout the water column may be typical. Thus, one possible explanation for the less dominant subsurface chlorophyll peak and the drop in water column primary production observed at the buoy station in mid-July may be that there was at this time no longer an overlap between the euphotic zone and the nutrient-rich bottom water.

It should, however, be noted that this explanation assumes that the subsurface peak is being maintained with bottom nutrients and, thus, comprises a large percentage of "new production" (high "f-ratio", i.e. % of total production comprised by new production as defined by Epply and Peterson [1979]). Assuming that these peaks are maintained by algal growth in situ (i.e. in the subsurface peak) and not by advection, the relatively large standing stocks observed would also support this argument. Nevertheless, the work of Kaas et al. [1992], in which nitrate uptake was examined using ^{15}N tracer techniques, indicates that the subsurface phytoplankton peak may on some occasions exhibit a relatively low f-ratio.

Kaas et al.'s argument is, in part, based on a hydrographic analysis of the conditions at the buoy station [Pedersen, 1991] which suggests that the subsurface phytoplankton peak may often be associated with an "intrusion" layer (i.e. water with temperature and salinity characteristics which differ from both the surface and bottom waters). Phytoplankton confined to this intrusion layer would not, under normal conditions, have immediate access to nutrients in the bottom waters. Further work is clearly necessary to describe the fate of the carbon fixed by subsurface chlorophyll peaks in this region.

5.8. Mesozooplankton Response to Phytoplankton Distribution/Activity

The production (eggs produced per female per day) of two common copepods and the biomass distribution of the combined copepod populations throughout the year at the buoy station are illustrated in Figure 5.8. In addition, the surface water temperature and the integrated water column biomass of phytoplankton >11 μm (i.e. those above the minimum grazable size for most copepods) are also shown.

Copepod egg production responded clearly to the periods of the spring bloom and the autumnal breakdown of stratification. The marked reaction of copepod production to what are presumed to be peaks in new production (as evidenced by the relatively large numbers of net (i.e. large) phytoplankton observed during the spring bloom and at the time of the autumnal breakdown of stratification (see Kiørboe [1993]) makes the measurement of copepod egg production uniquely suited as an indicator of changes in nutrient availability in the pelagic food web.

In contrast, total copepod biomass at this station in the Kattegat showed a strong correlation with water temperature and exhibited a single peak during mid-summer. The fact that biomass showed little response to peaks in primary production is a function of the relatively long generation time of copepods and the temperature sensitivity of the generation time [e.g., McLaren, 1978].

5.9. Short-term Fluctuations in Carbon Flow

The effect of wind-generated mixing of nitrate-rich bottom water on total primary production and on the structure of the pelagic food web was studied during one of the intensive study periods at the buoy station (October 24 to November 9, 1988).

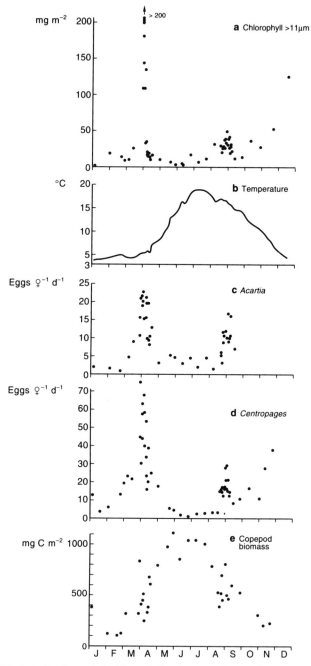

Figure 5.8. Distribution of a) chlorophyll-*a* integrated over the water column, b) temperature, c) egg production per *Acartia* female, and d) egg production per *Centropagus* female and e) copepod biomass at the buoy station. From Kiørboe and Nielsen [1994].

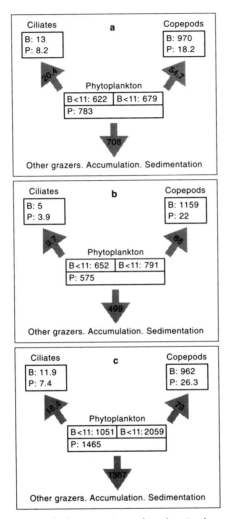

Figure 5.9. Carbon budgets at the buoy station referred to in the text a) before, b) during, and c) after a storm in October–November, 1988. Units for biomass (B) and production (P) are mg C m^{-2} and mg C m^{-2} d^{-1}, respectively. From Nielsen and Kiørboe [1991]. Reprinted by permission of Oxford University Press.

That study is described in detail by Kiørboe and Nielsen [1990], and Nielsen and Kiørboe [1991]. Strong winds during October had gradually eroded the pycnocline and a storm with winds of up to 25 m s^{-1} at the beginning of November resulted in a water column which was mixed almost to the bottom. The mixing of the water column led to an input of >35 mmol nitrate m^{-2} to the photic zone at the buoy position during the storm. This nutrient input gave rise to an increase during the days following the storm in phytoplankton biomass and primary production (Figures 5.8 and 5.9) which was most pronounced for the larger (>11 μm) phytoplankton species.

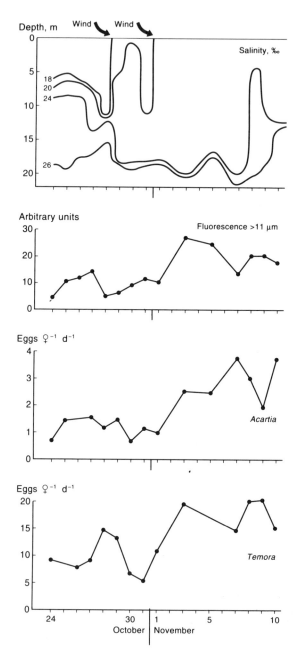

Figure 5.10. Temporal variation in the vertical distribution of a) salinity (simplified), b) fluorescence in the >11-μm fraction at 2.5 m, c) fecundity of *Acartia clausii*, and d) fecundity of *Temora longicornis*. From Kiørboe and Nielsen [1990].

This increase in large phytoplankton species resulted in a statistically significant increase (doubling) of copepod production measured as egg production per female (Figure 5.10). The increased egg production gave rise to nearly a doubling in the concentration of copepod nauplii. However, as in the annual case described above, the copepod biomass remained essentially unchanged in the period following the storm (data not shown). Again, this is because nauplii do not contribute greatly to the total copepod biomass.

This example illustrates the potential for substantial changes in the pattern of carbon flow in the pelagic community over very short times scales in this region in response to changing weather (wind) conditions. Appreciation of the importance of local weather conditions for the flow of energy in the pelagic ecosystem has only recently emerged (see Kiørboe, [1993] and chapter 4, this volume) and there is a need for further study in order to quantify the relationship between wind conditions and energy flow as well as the interannual variability in production processes that can be ascribed to differing weather conditions.

5.10. Long-term Changes in Organic Production

A prerequisite in order to quantify changes in organic production over time is the existence of a good historical data set describing primary production. The most sensitive and common method used today to determine primary productivity in aquatic environments was first described in 1952 by Steemann Nielsen. The method was rather a novelty in that it requires the addition of radioactively labeled CO_2 and quite a number of years passed before it was routinely used. Thus, for most marine regions, there are no historical primary production data from before about the mid-1960s which is comparable to those collected today.

Fortunately for this study, Steemann Nielsen, himself, carried out a comprehensive study with the aim of describing annual primary production in the Kattegat during the period 1954–1960. In his study [Steemann Nielsen, 1964], samples were taken at approximately fortnightly intervals at a fixed station in the northern Kattegat (56°51'2 N; 11°48'3 E) and particulate primary production was determined following in situ incubations. Steemann Nielsen's conclusion was that the annual primary production during this period was between 51 and 82 g C m^{-2} yr^{-1} (mean = 67).

No attempt has been made to exactly duplicate Steemann Nielsen's methods and sampling strategy. However, Richardson and Heilmann [1995] have collated all available particulate primary production data collected during the period 1984–1993 from stations near the fixed position of Steemann Nielsen. In so far as possible, they attempted to apply the same methods of data treatment as Steemann Nielsen. Using these procedures, they conclude that annual primary production in the Kattegat in their study period was on the order of 240 g C m^{-2} yr^{-1}.

Their analysis indicated that primary production estimates during the winter months (when light is predicted to be limiting for primary production) are similar during the 1950s and the years between 1984 and 1993. However, from about the

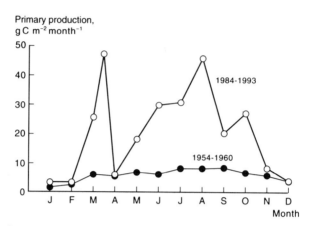

Figure 5.11. Daily primary production at four different depths in the water column throughout the year as estimated by Steemann Nielsen [1964] (closed circles) and Richardson and Heilmann [1995] open circles.

time of the spring bloom and for most of the months during the summer (when nutrient availability is predicted to control primary production) estimates of primary production are very much higher in the 1984–1993 study period than in the 1950s (Figure 5.11). This observation led these authors to conclude that the apparent difference observed in annual primary production between the 1950s and the 1980s is a consequence of eutrophication (see chapter 3 for a discussion of the extent of the eutrophication occurring in this period).

5.11. Conclusions

The comprehensive studies of energy flow in the pelagic community in the Kattegat carried out as a part of the Danish Marine Research Programme have indicated that there has, indeed, been a substantial increase in the magnitude of the primary production occurring annually in this semienclosed sea since the 1950s – a period in which land-based sources of nitrogen to this semienclosed sea have approximately doubled [Hansen et al., 1995]. Recorded sedimentation rates combined with analyses of advection rates of oxygen into the bottom waters of the Kattegat and through the pycnocline support the hypothesis that the increased frequency and intensity of hypoxia and anoxia observed in this region can be directly related to the primary production occurring locally.

Sedimentation of particulate organic carbon was not, as expected, largely confined to the period during and immediately following the spring bloom but continued at a relatively constant level throughout the summer months. Carbon flow within the pelagic food web is highly dynamic and seemingly small changes in wind direction and/or intensity can give rise to nutrient transfer from bottom to surface waters in this stratified sea which can, in turn, dramatically alter energy flow.

References

Ærtebjerg, G., Causes and effects of eutrophication in the Kattegat and the Belt Sea (in Danish), in *Eutrofiering av havs- och kustområden*, 22, Nordiska Symposiet om Vattenforskning, Laugarvatn 1986. Nordforski, Miljövårsserien, 1987.

Ærtebjerg, G., Marine areas, fjords, coasts, and the open sea (in Danish), *The Water Quality Act's Monitoring Programme 1991, Special Report 61*, Danish Environmental Protection Agency, 1992.

Bjørnsen, P., L. Hansen, B. Løkkegaard, and L. Berg, Bacteria plankton and bacteria grazing (in Danish), in *Plankton Dynamics and Carbon and Nutrient Flow in Kattegat*, edited by T. Fenchel, *Havforskning fra Miljøstyrelsen*, 10, 137–150, Danish Environmental Protection Agency, Copenhagen, 10, 1992.

Dugdale, R. C., and J. J. Goering, Uptake of new and regenerated forms of nitrogen in primary productivity, *Limnol. Oceanogr.*, 12, 196–206, 1967.

Epply, R. W., and B. J. Peterson, Particulate organic matter flux and planktonic new production in the deep ocean, *Nature*, 282, 677–680, 1979.

Granéli, E., Nutrient limitation of phytoplankton biomass in a brackish water bay highly influenced by river discharge, *Estuar. Coast. Shelf. Sci.*, 25, 563–569, 1987.

Hansen, I. S., G. Ærtebjerg, L. A. Jørgensen, Analysis of oxygen sedimentation in the Kattegat, the Belt Seas and the western Baltic (in Danish), *Havforskning fra Miljøstyrelsen*, 1, 133 pp., Danish Environmental Protection Agency, Copenhagen, 1990.

Hansen, I. S., G. Ærtebjerg, K. Richardson, J. P. Heilmann, O. V. Olsen, and F. B. Pedersen, Effects of reduced nitrogen input on oxygen conditions in the inner Danish waters (in Danish), *Havforskning fra Miljøstyrelsen*, 29, Danish Environmental Protection Agency, Copenhagen, 103 pp., 1994.

Hansen, I. S., G. Ærtebjerg, and K. Richardson, A scenario analysis of effects of reduced nitrogen input on oxygen conditions in the Kattegat and the Belt Sea, *Ophelia*, 42, 75–93, 1995.

Hansen, P. J., and T. G. Nielsen, Microzooplankton (in Danish), in *Plankton Dynamics and Carbon and Nutrient Flow in Kattegat*, edited by T. Fenchel, *Havforskning fra Miljøstyrelsen*, 10, 61–76, Danish Environmental Protection Agency, Copenhagen, 1992.

Hansen, P. J., Quantitative importance and trophic role of heterotrophic dinoflagellates in a coastal pelagical food web, *Mar. Ecol. Prog. Ser.*, 73, 253–261, 1991.

Heath, M., K. Richardson, and T. Kiørboe, Optical assessment of phytoplankton nutrient depletion, *J. Plank. Res.*, 12, 381–396, 1990.

Heilmann, J. P., K. Richardson, and G. Ærtebjerg, Annual distribution and activity of phytoplankton in the Skagerrak-Kattegat frontal region, *Mar. Ecol. Prog. Ser.*, 112, 213–223, 1994.

Jespersen, A. M., M. Søndergaard, K. Richardson, and B. Riemann, Estimate of the accuracy and precision of the ^{14}C-method for determination of the primary production of planktonic algae by use of routine methodology (in Danish), *Havforskning fra Miljøstyrelsen*, 55, Danish Environmental Protection Agency, Copenhagen, 53 pp., 1995.

Jørgensen, B. B., and N. P. Revsbech, Oxygen uptake, bacterial distribution, and carbon-nitrogen-sulfur cycling in sediments from the Baltic Sea-North Sea transition, *Ophelia*, 31(1), 29–49, 1989.

Kaas, H., H. H. Kaas, and F. Møhlenberg, Upward directed transport of nutrients salts to the photic zone (in Danish), in *Plankton Dynamics and Carbon and Nutrient Flow in Kattegat*, edited by T. Fenchel, *Havforskning fra Miljøstyrelsen*, 10, 121–136, Danish Environmental Protection Agency, Copenhagen, 1992.

Kiørboe, T., and T. G. Nielsen, Effects of wind stress on vertical water column structure, phytoplankton growth, and productivity of planktonic copepods, in *Trophic Relationships in the Marine Environment*, edited by M. Barnes, and R. N. Gibson, pp. 28–40, Aberdeen University Press, 1990.

Kiørboe, T., and T. G. Nielsen, Mesozooplankton, production and grazing (in Danish), *Havforskning fra Miljøstyrelsen*, 10, 77–101, Danish Environmental Protection Agency, Copenhagen, 1992.

Kiørboe, T., Turbulence, phytoplankton cell size, and the structure of pelagic food webs, *Adv. Mar. Biol.*, 29, 1–72, 1993.

Kiørboe, T., and T. G. Nielsen, Regulation of zooplankton biomass and production in a temperate, coastal ecosystem. 1. Copepods. *Limnol. Oceanogr.*, 39, 493–507, 1994.

Kronvang, B., G. Ærtebjerg, R. Grant, P. Kristensen, M. Hovmand, and J. Kirkegaard, Nationwide monitoring of nutrients and their ecological effects: State of the Danish Aquatic Environment, *Ambio*, 22(4), 1993.

McLaren, I. A., Generation lengths of some temperate copepods: Estimation, prediction, and implications, *J. Fish. Res. Bd Can.*, 35, 1330–1342, 1978.

Nielsen, T. G., and T. Kiørboe, Effects of a storm event on the structure of the pelagic food web with special emphasis on planktonic ciliates, *J. Plank. Res.*, 13, 35–51, 1991.

Nielsen, T. G., and T. Kiørboe, Regulation of zooplankton biomass and production in a temperate, coastal ecosystem. 2. Ciliates, *Limnol. Oceanogr.*, 39, 508–519, 1994.

Nixon, S. W., Quantifying the relationship between nitrogen input and the productivity of marine ecosystems, *Nixon Pro. Adv. Mar. Tech. Conf.*, 5, 57–83, 1992.

Olesen, M., The fate of an early spring bloom in the Kattegat, *Ophelia*, 37(1), 51–66, 1993.

Olesen, M., Comparison of the sedimentation of a diatom spring bloom and of a subsurface chlorophyll maximum, *Marine Biology*, 121, 541–547, 1995.

Olesen, M., and C. Lundsgaard, Sedimentation of organic material from the photic zone in the southern Kattegat (in Danish), *Havforskning fra Miljøstyrelsen*, 10, 167–183, Danish Environmental Protection Agency, Copenhagen, 1992.

Olesen, M., and C. Lundsgaard, Seasonal sedimentation of autochthonous material from the euphotic zone of a coastal system, *Estuar. Coast. Shelf Sci.*, 1995, in press.

Pedersen, F. Bo, Hydrographic conditions in the southern Kattegat (in Danish), *Havforskning fra Miljøstyrelsen*, 3, Danish Environmental Protection Agency, Copenhagen, 1991.

Peterson, B. J., Aquatic primary productivity and the ^{14}C-CO$_2$ method: a history of the productivity problem, *Ann. Rev. Ecol. Syst.*, 11, 369–385, 1980.

Pettersson, H., Scattering and extinction of light in sea-water, Meddelanden Göteborgs Högskolas Oceanografiska Institution., 4b(4), 1934.

Planlægningsrådet for Forskningen, Nitrogen and phosphorus in the water environment: consensus report (in Danish), Planlægningsrådet for Forskningen, Forskningssekretariatet, Copenhagen, Denmark, 1987.

Redfield, A. C., The biological control of chemical factors in the environment, *Am. Sci.*, 46, 205–222, 1958.

Richardson, K., and A. Christoffersen, Seasonal distribution and production of phytoplankton in the southern Kattegat, *Mar. Ecol. Prog. Ser.*, 78, 217–227, 1991.

Richardson, K., and J. P. Heilmann, Primary production in the Kattegat: Past and present, *Ophelia*, 41, 317–328, 1995.

Rosenberg, R., R. Elmgren, S. Fleischer, P. Jonsson, G. Persson, and H. Dahlin, Marine eutrophication case studies in Sweden, *Ambio*, 19(3), 102–108, 1990.

Steemann Nielsen, E., The use of radio-active carbon (C[14]) for measuring organic production in the sea, *J. Cons. Intl Explor. Mer.*, 18(2), 117–140, 1952.

Steemann Nielsen, E., Investigations of the rate of primary production at two Danish light ships in the transition area between the North Sea and the Baltic, *Meddr Danm. Fisk.-Havunders.*, 4(3), 31–77, 1964.

Thomsen, H. (Ed.), *Plankton in the Inner Danish Waters* (in Danish), *Havforskning fra Miljøstyrelsen*, 11, 331 pp., Danish Environmental Protection Agency, Copenhagen, 1992.

6

Material Flux in the Sediment

Bo Barker Jørgensen

6.1. Introduction

Marine sediments accumulate in the coastal zone during interglacial periods, i.e. over the last ca. 6000 years since the sea reached its present level, and erode during glaciations. About 90% of the global flux of riverine particulates and organic carbon is currently trapped in deltaic and shelf regions [Gibbs, 1981]. Due to the fertilization by rivers, by coastal upwelling, or by intrusion of deep slope water onto the shelf, about 30% of the oceanic primary production takes place within the 10% of the ocean area which encompasses shelf and coastal systems. Some 25–50% of the organic carbon, nitrogen, and phosphorus produced by the local plankton communities sinks out of the water column to the sediment [Wollast, 1991]. Most of the deposited organic matter is, however, again mineralized and only about 10% of it becomes more permanently buried and contributes to the mean organic content of 1–3% dry weight in shelf sediments [Berner, 1982].

Due to their large storage capacity for organic matter and nutrients, the sediments have an important regulatory and buffering function in the coastal ecosystem. They affect the seasonal oxygen balance of the bottom waters and, through the release of nutrients to the water column, also affect the seasonal phytoplankton production. Due to the high rate of microbial metabolism in coastal sediments, they constitute an anoxic world where only a thin layer beneath the surface and around animal burrows contains oxygen. The depth of the oxic-anoxic interface changes over the year, depending on the deposition rate of organic carbon and the oxygen concentration in the overlying water. The sediments and the benthic communities are, therefore, the most sensitive parts of the coastal ecosystem to eutrophication and hypoxia. When the oxygen supply falls or the demand increases, dramatic changes occur in the benthos, in the dominant metabolic pathways of microbial processes,

Eutrophication in Coastal Marine Ecosystems
Coastal and Estuarine Studies, Volume 52, Pages 115–135
Copyright 1996 by the American Geophysical Union

in the sediment chemistry, and in the nutrient flux to the overlying water. These phenomena and their dependence on eutrophication are discussed in this chapter, using mostly data from Danish waters.

6.2. Sediment Types

The sea floor of the complex estuarine system extending from the Baltic Sea into the North Sea (Figure 6.1) developed as a submerged glacial landscape, shaped by the extensive ice covers more than 20,000 years ago. Meltwater deposits are still exposed in some areas. Extensive fine-sand sediments occur in the North Sea and in the western Kattegat where mud accumulation is prevented by currents and especially by wave action reaching down to the rather shallow bottom at depths of ca. 30 and 20 m, respectively. Mud accumulation, at a mean rate of ca. 500 g m^{-2} yr^{-1}, occurs mostly in deeper areas or in protected basins along the coasts, such as in the Belts and the western Baltic [Madsen and Larsen, 1986]. On the southern and western slopes of the deep, glacially eroded trough in the Skagerrak and northeast Kattegat, the sediment builds up at high rates of 2–4 mm yr^{-1} [van Weering et al., 1987].

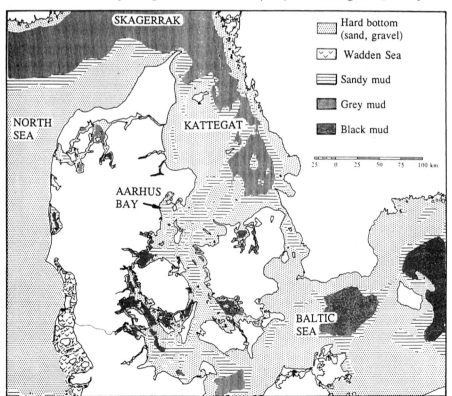

Figure 6.1. Sediment type distribution from the Baltic Sea to the North Sea. [From Larsen, 1968].

In a depositional fan in the northern Kattegat up to 10 mm yr^{-1} of mud accumulation has been measured. The organic sedimentation rate, 230 g org. C m^{-2} yr^{-1}, is here of the same magnitude as the local phytoplankton productivity (cf. chapter 5). The area covers only 10% of the area of inner Danish waters but accumulates 40% of the total mud sediments and 35% of the organic matter [Madsen and Larsen, 1986]. The main sources of the fine-grained sediment are the large North European rivers discharging into the southeast North Sea. From there, much of the material is carried into the Skagerrak and Kattegat by the Jutland Current, after many sedimentation–resuspension cycles in the North Sea [e.g., Eisma and Kalf, 1987]. Some 40–50% of the organic carbon, which is deposited on the sea floor in the northern Kattegat area, also accumulates there. This high fraction is explained by the poor degradability of the organic material after the long transport from the North Sea [Jørgensen et al., 1990]. In other areas of Danish waters, only 10–15% of the deposited org. C resists mineralization before burial below 1–2 m of sediment.

These burial fractions correspond well to the average values for coastal sediments and also reflect the general trend of higher burial efficiency of organic matter at higher sedimentation rates [Henrichs and Reeburgh, 1987]. The burial efficiency of organic carbon and nutrients is generally higher in the absence of oxygen and of bottom fauna. This has been demonstrated most clearly for permanently anoxic (euxinic) basins [Canfield, 1989], whereas the effect of only intermittent hypoxia, which occurs in Danish coastal waters, still remains to be demonstrated. Benthic animals are important for the mechanical degradation of detritus and for the maintenance of a mixed sediment zone, mostly 3–6 cm deep [Madsen and Larsen, 1986; van Weering et al., 1987; chapter 7, this vol.], in which the organic matter is exposed to oxygen at intervals.

The degradation of organic matter continues at decreasing rates deep down in the sediment, even into the sulfate-depleted zone below several meters depth, where the microbial metabolism leads to methane formation. As a result, the methane pressure builds up in the sediment of many of the Danish coastal basins to above the ambient hydrostatic pressure. Thus, the methane forms gas bubbles trapped in the sediment, which can be seen as acoustic scattering layers at 2–5 m depth in seismic profiles. Methane from deeper sources in the northern Kattegat and Skagerrak sediments also escapes as bubbles into the water column and creates "pock marks" at the sediment surface or causes the formation of carbonate chimneys as a result of methane oxidation to CO_2 within the sediment [Jensen et al., 1992; Dando et al., 1994]. These sediment localities harbor a special and interesting fauna of pogonophores and bivalves with symbiotic methane- or sulfide-oxidizing bacteria in their tissue.

6.3. Deposition of Organic Matter

The main source of organic matter to estuarine and shelf sediments is the deposition of detrital material from the local plankton community in the overlying water. Consequently, there is a rough correlation between the phytoplankton primary productivity, the water depth through which the detritus sinks under progressing decomposition, and the primary sedimentation on the sea floor [e.g., Suess, 1980; Hargrave, 1984]. There are, however, many factors which complicate and delay the coupling of these processes to the benthic metabolism, including the seasonal temperature, the degradability and decay time of the organic matter, the burial of organic matter into anoxic sediment layers, or the resuspension and lateral transport of organic matter to deeper or less exposed areas.

As an example, organic-rich ephemeral mud blankets in Laholm Bay on the Swedish Kattegat coast were observed to form on the more sandy sediment surface after the sedimentation of the spring phytoplankton bloom [Floderus and Håkanson, 1989]. During successive resuspension cycles, these mud blankets were transported out of the bay and into the deeper parts of the Kattegat. Lateral advection was similarly found to redistribute the deposited organic matter toward the deeper parts of the Kiel Bight in the western Baltic Sea ("sediment focusing" [Balzer et al., 1986; Graf, 1992]). The distribution of sandy and muddy sediments in Figure 6.1 shows areas of relative export and import, respectively, of organic material through lateral advection of organic-rich sediment. Although the organic material may be relatively refractory after many resuspension cycles, it potentially enhances the local oxygen demand and nutrient release within the sedimentation basins. These basins tend to have organic-rich sediments and develop hypoxia in the poorly ventilated bottom water.

The flux of organic matter to the sediment has been measured by sediment traps suspended at different depths in the water column. It is difficult in shallow coastal systems to determine the net flux of primary sedimenting material on the background of a much larger gross flux of resuspended sediment. A discrimination between primary and secondary sedimentation has been made in some studies by the use of fresh chlorophyll or other chemical constituents for correction [e.g., Kemp and Boynton, 1992]. Due to the salinity and temperature stratification of the Baltic Sea and Kattegat waters (chapter 3), the upward transport of resuspended sediment is most of the year blocked by the pycnocline, and sediment trapped above this pycnocline is, therefore, representative of the net deposition.

In a field study in the Aarhus Bay (chapter 7) this blocking effect was used to discriminate seasonal gross and net deposition rates (Pejrup et al., submitted). The data in Figure 6.2 show a halo- and thermocline at 8–13 m depth, below which the amount of particulate material collected by the traps increased exponentially down toward the sediment surface at 16 m depth. By extrapolation of sediment trap data from 0–11 m down to 16 m, a net deposition of 2.9 g m^{-2} d^{-1} was determined. Due to frequent resuspension, the gross deposition was 30-fold higher, even though the data period was a relatively calm summer with a well-developed pycnocline.

The fraction of the primary production in coastal waters, which is deposited to the sediment, is mostly in the range of 25–50% of the primary productivity [Nixon, 1981; Wollast, 1991]. With a productivity of eutrophic coastal waters such as the Kattegat of 250–300 g C m^{-2} yr^{-1}, the mean organic carbon flux to the sediment should be in the range of 75–150 g C m^{-2} yr^{-1}. The annually deposited fraction

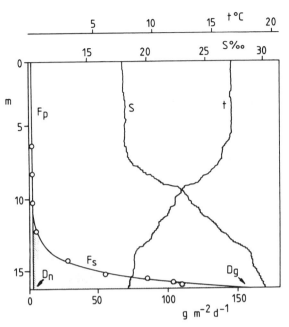

Figure 6.2. Sediment trap data: Total dry weight of sediment (○), trapped at different water depths during July 2–26, 1990, in central Aarhus Bay. F_p = flux of primary sediment; F_s = total downward flux of sediment; D_n = net deposition rate; D_g = gross deposition rate. (Data from Valeur et al. [1992]). Salinity and temperature stratification measured on June 21, 1990, demonstrates a pycnocline at 8–13 m depth. (Aarhus Amtskommune, unpublished report).

in the 10–30 m deep Baltic and Kattegat waters was: 26% in southern Kattegat (chapter 5), 32–45% in Aarhus Bay (chapter 7), and about 30% in the Kiel Bight [Smetacek, 1980].

The fraction of the phytoplankton production which sinks out of the euphotic zone may vary strongly over the year. There are many examples of the rapid deposition of the spring bloom of diatoms which, upon nutrient depletion or due to wind-driven turbulence, suddenly aggregate and over a few days or weeks sink to the bottom (chapters 4 and 5). The larger the bloom and the faster it develops, the smaller a fraction appears to be grazed or mineralized in the water column, and the more is deposited [e.g., Smetacek, 1985]. This was also the case in the Kattegat where, however, the spring bloom accounted for only 8% of the yearly net carbon flux to the sediment and sedimentation was found to proceed at a steady rate throughout the phytoplankton growth season [Olesen and Lundsgaard, 1995]. Due to the sudden deposition of the spring bloom and of blooms, which may develop later during the summer just below the pycnocline, bursts of bacterial growth and metabolic activity occur at the sediment surface [Graf et al., 1983; Meyer-Reil, 1983]. These bursts appear to fade out again within 2–4 weeks as the most labile organic material is depleted and the rest of it becomes buried into the sediment [e.g., Hansen and Blackburn, 1992; chapter 7, this vol.].

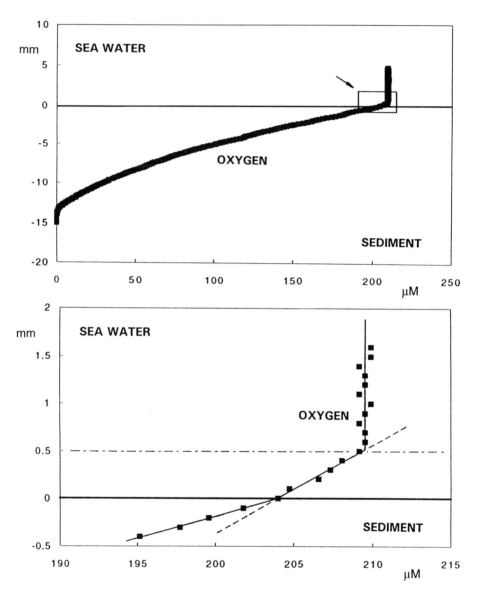

Figure 6.3. Oxygen distribution across the sediment-water interface measured in situ at 600 m water depth in central Skagerrak by a microelectrode mounted on a benthic lander, "Profilur". The data are presented at two different resolutions to demonstrate (top) the sharp oxic-anoxic interface on a macroscopic scale and (bottom) the occurrence of a 500 μm thick diffusive boundary layer in which the 100-μm depth intervals of oxygen measurements are just sufficient to allow accurate diffusion flux calculations. [Gundersen et al., 1994].

The contribution of terrestrially derived organic material to coastal waters is generally poorly known but is estimated to be small relative to the plankton material. Thus, although both lignin and polysaccharide components of vascular plants were identified in sediment trap material from a semienclosed bay on the Pacific Coast of Washington, USA, they together comprised <10–35% of the total sedimenting organic carbon [Hedges et al., 1988]. The contribution of particulate organic material from land to the Kattegat was estimated to be in average 8 g C m^{-2} yr^{-1} equivalent to 3% of the primary production [Olesen and Lundsgaard, 1995].

6.4. The Sediment-Water Interface

The sedimenting organic material is primarily mineralized by aerobic microorganisms and animals on the sea floor. The *oxic* (i.e. oxygen-containing) zone of coastal sediments is, however, generally restricted to a few millimeters thin layer at the sediment surface and surrounding the burrows of macrofauna [Revsbech et al., 1980; Jørgensen and Revsbech, 1985]. This oxic layer comprises only the uppermost part, often only 10–20%, of the brown (*oxidized*) sediment layer [Jørgensen and Revsbech, 1989]. Although the oxygen uptake is regulated overall by the net deposition and degradability of organic matter, the mechanism and dynamics of the oxygen uptake are complex. Oxygen penetrates into nonpermeable sediments mostly by molecular diffusion through the diffusive boundary layer, which constitutes an 0.2–1 mm thick, unstirred water film overlying the sediment. This boundary layer is apparent from high resolution measurements with oxygen microelectrodes applied directly on the sea floor [Gundersen and Jørgensen, 1990]. Figure 6.3 shows an example from the deep part of the Skagerrak, at 600 m water depth, where oxygen penetrated 13 mm into the sediment.

By diffusion flux calculations from boundary layer oxygen gradients or by diffusion-reaction modeling of the whole oxygen profile in the sediment, the respiration rate in the oxic surface layer can be calculated (Figure 6.4). During summer, there is often a maximum in respiration rate at the lower boundary of the oxic zone where reduced compounds such as ammonium diffuse up from below and are oxidized, e.g. by nitrifying bacteria. In sediments of the southern Kattegat and Aarhus Bay, it was found that the diffusive uptake accounted for 45–65% and 70%, respectively, of the total oxygen consumption measured in sediment cores [Rasmussen and Jørgensen, 1992]. The unaccounted uptake is mostly due to biopumping and respiration of burrowing fauna and the relative difference between diffusive and total uptake is, accordingly, closely related to the density of benthic animals [Glud et al., 1994; cf. chapter 8, this vol.]. Furthermore, the interaction between the overflowing sea water and sediment surface topography creates lateral pressure gradients which, in porous sediments, may drive an advective pore water transport through the sediment and, thereby, enhance the oxygen consumption and exchange of nutrients [Hüttel and Gust, 1992].

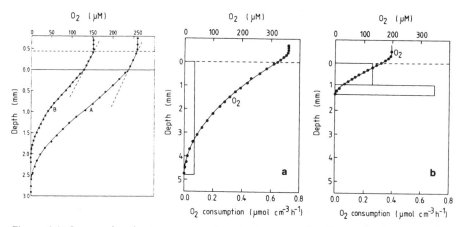

Figure 6.4. Oxygen distributions measured in situ in a coastal sediment (Aarhus Bay, 16 m water depth). Left: Oxygen gradients in the diffusive boundary layer are used to calculate the diffusive flux into the sediment as the simple product of gradient and diffusion coefficient. Center and right: Oxygen gradients before (a) and after (b) the deposition of a spring phytoplankton bloom in late March. The thickness of the oxic zone varied from 4.8 mm (a) to 1.2 mm (b) over the year. The rate of respiration in the oxic zone was calculated per unit sediment volume by a zero-order diffusion reaction model. Before the sedimentation event, the rate of respiration was uniform throughout the oxic zone. After the event, the respiration in the much thinner oxic zone had increased several-fold, with a maximum at the oxic-anoxic interface at 1.2 mm depth. The calculated area oxygen uptake rate increased from 265 (a) to 550 (b) μmol O_2 m^{-2} h^{-1} following the sedimentation event. From Rasmussen and Jørgensen [1992].

6.5. Anaerobic Mineralization Processes

The degration and oxidition of the organic matter below the thin oxic zone is mostly carried out by microorganisms. Some of these organisms hydrolyze and ferment the macromolecules to small organic compounds, while others use oxidants other than O_2 to respire the compounds to CO_2. Although the net result of carbon oxidation may be similar, there are major differences between the anaerobic and the aerobic food chains, in which the organisms basically have a very uniform energy metabolism, the aerobic respiration. The anaerobic, microbial food chain comprises a depth sequence of metabolic types, each of which is only able to exploit a fraction of the energy available in the organic matter and each of which excretes energy-rich inorganic or organic products, which can again be oxidized when they reach up into the overlying chemical zones. Table 6.1 shows a range of such redox reactions which may be catalyzed by bacteria. Some of these may also proceed chemically, e.g. the oxidation of Mn^{2+}, Fe^{2+} or H_2S by O_2.

There is a more or less distinct vertical zonation of the dominant energy metabolisms in coastal sediments, with a) aerobic respiration in the uppermost few mm, b) nitrate respiration (denitrification) in the following few mm-cm, c) manganese

TABLE 6.1. Mineralization processes in marine sediments by which organic matter becomes oxidized with different electron acceptors. The reoxidation of the reduced products with oxygen is also shown. After Canfield et al. [1993].

O_2 respiration:
$$CH_2O + O_2 \rightarrow CO_2 + H_2O$$

Denitrification:
$$CH_2O + {}^4/_5NO_3^- + {}^4/_5H^+ \rightarrow CO_2 + {}^2/_5N_2 + {}^7/_5H_2O$$

Mn reduction:
$$CH_2O + 2MnO_2 + 4H^+ \rightarrow 2Mn^{2+} + 3H_2O + CO_2$$
$$2Mn^{2+} + O_2 + 2H_2O \rightarrow 2MnO_2 + 4H^+$$

$$CH_2O + O_2 \rightarrow CO_2 + H_2O$$

Fe reduction:
$$CH_2O + 4FeOOH + 8H^+ \rightarrow 4Fe^{2+} + CO_2 + 7H_2O$$
$$4Fe^{2+} + O_2 + 6H_2O \rightarrow 4FeOOH + 8H^+$$

$$CH_2O + O_2 \rightarrow CO_2 + H_2O$$

Sulfate reduction:
$$CH_2O + {}^1/_2SO_4^{2-} + H^+ \rightarrow CO_2 + {}^1/_2 H_2S + H_2O$$
$${}^1/_2H_2S + O_2 \rightarrow {}^1/_2SO_4^{2-} + H^+$$

$$CH_2O + O_2 \rightarrow CO_2 + H_2O$$

Methanogenesis:
$$CH_2O \rightarrow {}^1/_2CH_4 + {}^1/_2CO_2$$
$${}^1/_2CH_4 + O_2 \rightarrow {}^1/_2CO_2 + H_2O$$

$$CH_2O + O_2 \rightarrow CO_2 + H_2O$$

NH_4^+ oxidation:
$$NH_4^+ + 2O_2 \rightarrow NO_3^- + H_2O + 2H^+$$

(Mn(IV)) and d) iron (Fe(III)) reduction in the main part of the brown oxidized sediment, which may be a few to ten centimeters deep, e) sulfate reduction, which mostly extends as the dominant process down to one or several meters depth, and f) methanogenesis, which is the terminal metabolism for organic matter after depletion of the mentioned oxidants. Denitrification is in this context exceptional in that the reduced product, N_2, has a stable triple-bond, which renders it useless as an energy substrate, even in the presence of oxygen. Ammonium released by the anaerobic degradation of nitrogen-containing organic compounds is, however, an excellent energy substrate used by the nitrifying bacteria. Although nitrate may be reduced directly to ammonium by several types of anaerobic bacteria, the process is of much less significance than denitrification in marine sediments [e.g., Binnerup et al., 1992].

Denitrification is of major importance for the nitrogen cycle and nutrient balance in estuaries and may reach high rates under certain conditions, such as high nitrate

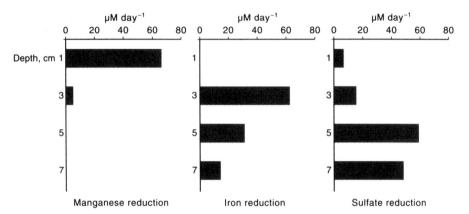

Figure 6.5. Distributions of manganese, iron, and sulfate reduction rates in sediment from northern Kattegat. The Mn(IV) and Fe(III) reduction rates were measured from the accumulation of Mn^{2+} and Fe^{2+} in the presence of molybdate to inhibit sulfate reduction. Sulfate reduction was measured in a separate incubation by radiotracer technique. From Canfield [1993]. Reprinted by permission of Springer-Verlag.

load or intensive biopumping (1–3 mmol NO_3^- m^{-2} d^{-1}, Sørensen et al., 1979; Seitzinger, 1988]. Nevertheless, the process was found to account for only a few percent of the organic matter oxidation in most Danish coastal sediments, with values increasing to 3–6% in the Skagerrak at 400–700 m depth [Canfield et al., 1993b; Nielsen et al., 1994]. Oxidized minerals of manganese and iron in the brown sediment layer can, however, be used as respiratory electron acceptors for the oxidation of organic carbon by certain bacteria [e.g., Lovley, 1987; Burdige, 1993]. Alternatively, they can serve as oxidants of sulfide produced in the sediment by bacteria respiring sulfate and, thus, contribute only indirectly to the oxidation of organic matter [Aller, 1994]. It is experimentally difficult to determine the rates of metal reductions or to discriminate their roles in organic carbon oxidation. Much information on processes and rates, therefore, derives from modeling of geochemical gradients in deeper sediments. By combinations of chemical and radiotracer methods it has, however, been possible to demonstrate the mutual zonation of the three processes in coastal sediments [Canfield et al., 1993a]. Although there is significant overlap (Figure 6.5), the depth distribution of the oxidation processes clearly follows a depth sequence of decreasing redox potential and energy yield. The sequence is probably determined by competition among the anaerobically respiring bacteria for common organic substrates [e.g., Lovley and Phillips, 1987].

The iron cycling is of particular interest, because Fe(III) minerals bind phosphate and thus regulate its release from the sediment. In addition, iron binds the large amounts of sulfide produced. Although manganese generally occurs in much lower concentrations than iron, Mn(IV) mineral phases have a faster turnover (chapter 7). Due to the much slower O_2 oxidation kinetics of reduced manganese than of reduced iron, Mn^{2+} may escape through the thin, oxic sediment zone and up into the

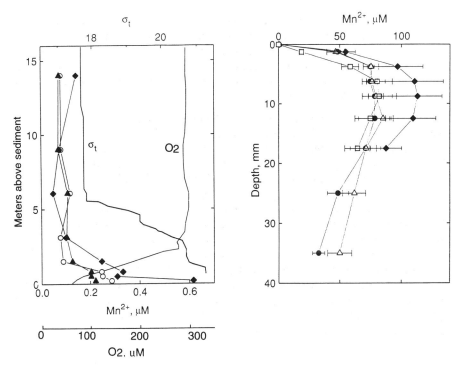

Figure 6.6. Distributions of dissolved manganese in the water column (left) and in the sediment pore water (right) in Aarhus Bay on three different days in August, 1992. The water density, σ_t, showed that a pycnocline created stable water stratification in the lowest 5 m, down through which the oxygen concentration dropped steeply toward the bottom (left). One third of the Mn^{2+} flux from the sediment was reoxidized near the sediment surface while two thirds diffused out of the sediment into the lower water column where it was oxidized to MnO_2 and subsequently redeposited. From Thamdrup et al. [1994], and Gundersen et al. [1994]. Reprinted with permission from Elsevier Science Ltd.

water column (Figure 6.6). There, it is subsequently oxidized to MnO_2 and is again deposited but at a different location due to lateral transport by bottom currents. Through hundreds of such emission-deposition cycles, manganese may gradually move away from the organic-rich, reduced coastal sediments to finally become trapped in more oxidized sediments on the continental slope. In the 700 m deep trough in central Skagerrak (see map in chapter 1), migrating manganese oxides are focused toward the deepest sediments in which manganese alone constitutes up to 5% of the dry weight in the upper 5 cm. Mn(IV) is here the main oxidant for the mineralization of organic matter [Canfield et al., 1993b]. At intermediate depths of around 400 m in the Skagerrak, Fe(III) was the most important oxidant.

In organic-rich coastal sediments, most anaerobic mineralization is by sulfate reduction, which may here be of similar importance and efficiency as aerobic respiration for the oxidation of organic carbon [Jørgensen, 1982; chapter 7, this vol.]. Sulfate respiration proceeds at steadily decreasing rates with depth down to the depletion of sulfate at several meters below the sediment surface, where the methanogenic bacteria then take over. The large amounts of H_2S formed by the sulfate reducers dominate the chemistry of the sediments in which black iron sulfides (FeS) and, especially, pyrite (FeS_2) accumulate. Calculated as potential oxygen demand, the burial of FeS_2 in these coastal sediments is equivalent to 10–25% of the burial of organic carbon [Jørgensen et al., 1990].

Depending on the availability of oxidants, the H_2S is continuously reoxidized in the sediment. The removal of sulfide is important for the benthic fauna due to the interference of H_2S with the aerobic respiratory system (chapter 8). The H_2S oxidation appears to take place mostly in the anoxic sediment through the partial oxidation and binding of sulfide by manganese and iron and the subsequent transport of iron sulfides up to the surface layers where O_2 is the terminal oxidant (chapter 7). Only at very high organic loading of the sediment or during hypoxia are the reactive Mn(IV) and Fe(III) pools depleted and H_2S may reach directly up to the oxic layer. There, sulfur bacteria carry out the oxidation in a narrow zone (ca. 0.1 mm) of overlapping, counter-diffusing O_2 and H_2S [Jørgensen and Revsbech, 1983]. In particular the filamentous *Beggiatoa* spp. proliferate and grow as sub-mm thin, white films on the FeS-blackened sediment surface. Due to eutrophication, these sulfur bacteria have spread during the last decade concurrently with the progressing hypoxia in North European coastal waters and they have even become a periodically dominant bottom community in some areas.

6.6. Nitrogen Cycling

Nitrogen and phosphorus are deposited to the sea floor bound in sedimenting algal cells and organic detritus, which in the water column undergo partial decomposition with a faster loss of the nutrients, N and P, than of carbon. When reaching the sediment, the organic material is, thus, partly depleted in N and P relative to the Redfield molar ratio of C:N = 6.5 and C:P = 106 in the plankton. Further mineralization in the sediment depletes the organic nitrogen to a C:N of >10. In the northern Kattegat, where extensive degradation of the organic matter has taken place during its transport from the North Sea, the C:N is even 15 at the sediment surface and increases to >20 with depth [Jørgensen et al., 1990]. In contrast, the C:P ratio is lowest at the sediment surface because released phosphate accumulates here, bound to iron minerals. The C:P ratios deeper in the sediment reach 300–600 and are generally maximal at sedimentation rates typical of shelf sediments, 0.1–1 mm yr^{-1} [Ingall and van Cappellen, 1990].

A low C:N ratio is indicative of relatively fresh, undegraded organic matter. Accordingly, the seasonal isopleths of C:N in a sediment from Aarhus Bay (Figure 6.7) show the deposition of the spring diatom bloom in two consecutive years and its rapid mixing down to 4–6 cm depth by bioturbation. A bloom of dinoflagellates was deposited in August (cf. chapter 7), while during the rest of the year the influx of nitrogen-rich material was lower and the sediment C:N remained rather constant.

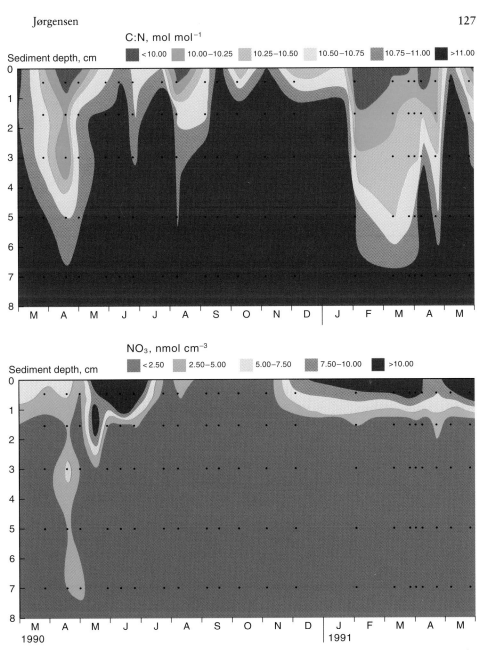

Figure 6.7. Isopleths of (top) C:N molar ratios of organic matter and (bottom) nitrate in the pore water in a marine sediment from Aarhus Bay during 1990–1991. From Lomstein and Blackburn [1992].

During the degradation of the N-containing compounds (mostly proteins and nucleic acids), urea $(CO(NH_2)_2)$, amino acids and other, still poorly defined dissolved organic nitrogen (DON) compounds are produced. In addition to ammonium, they are partly released to the overlying water. The urea flux appears to be of similar magnitude as the ammonium flux in several coastal sediments [Lomstein et al., 1989; chapter 7, this vol.]. Urea is excreted as a nitrogen waste product by the benthic fauna and is also produced by several anaerobic bacteria [Pedersen et al., 1993].

Some 50–75% of the ammonium produced was found to be nitrified to NO_3^- in different shelf sediments of the North Sea, the Kattegat, and the Bering Sea [Billen, 1978; Lomstein and Blackburn, 1992; Henriksen et al., 1993]. There is a complex balance between the a) flux of ammonium and DON into the water, b) nitrification of ammonium to nitrate, c) flux of nitrate into the water, and d) denitrification of nitrate to N_2. Thus, biopumping by the infauna enhances the advective ammonium flux, which may bypass the nitrification zone, but at the same time the biopumping brings oxygen down into the animal burrows and, thereby, expands the nitrification zone. Nitrification is also sensitive to eutrophication which may change the balance of oxygen and ammonium fluxes away from the optimum [Blackburn and Blackburn, 1992]. Nitrification may, thus, be strongly impeded by hypoxia and by high organic deposition with the result that ammonium is directly released into the water [Klump and Martens, 1987; Caffrey et al., 1993]. Since a large part of the denitrification is dependent on NO_3^- produced by nitrification, denitrification is also impeded, thereby removing less N nutrients from the ecosystem. This positive feedback mechanism thus tends to enhance the adverse effects of eutrophication [Kemp et al., 1990].

Denitrification takes place just below the O_2 zone and is limited by the availability of nitrate entering by diffusion or biopumping into the anoxic sediment. The zone of denitrification may be only 1–2 mm thick, similar to the O_2 zone, as shown for a Danish estuary by microelectrode technique [Binnerup et al., 1992]. Denitrification may also extend to one or several centimeters, deepest when the nitrate concentration in the overlying water is high and the microbial activity is relatively low (Figure 6.7). Due to the short diffusion path between the O_2 and NO_3^- zones, coupled nitrification-denitrification seems to dominate over nitrate coming from the water column in many coastal sediments. If, however, the bottom water concentration exceeds 30–50 μM NO_3^- or if nitrification is suppressed, e.g. by hypoxia, external nitrate becomes the more important source [Nielsen et al., 1994]. In spite of large seasonal and spatial variations in denitrification, it has been observed in Danish coastal sediments that several counteracting factors affecting this process lead to a rather constant denitrification rate over the year of around 400 μmol N m^{-2} d^{-1} [Nielsen et al., 1994].

The removal of combined nitrogen by denitrification in estuaries and costal environments is important for the nitrogen cycle and crucial for the understanding of the effects of eutrophication (see chapter 9). The fraction of the total nitrogen input to different coastal ecosystems which is removed again through this process has been estimated to range from 10% to >60% [e.g., Mantoura et al., 1991]. Although for a number of estuaries the estimated values fall mostly in the range of 30–50% [Seitzinger, 1988], the efficiency is often much lower and is strongly dependent on the retention time of the estuarine water mass [Nielsen et al., 1994]. The burial of organic nitrogen constitutes a smaller component in the N budget, being equivalent to about 1% of the annual N requirement for primary production in inner Danish waters [Jørgensen et al., 1990].

6.7. Phosphorus Cycling

Although phosphorus does not undergo reduction-oxidation processes in marine sediments, its chemistry is complicated by the diverse forms of binding in organic and inorganic phases, and much research effort has been devoted to methods for the identification of these pools. Much of the phosphorus deposited on the sediment is bound in fresh plankton material and is gradually released to the pore water during mineralization. The free phosphate is rapidly bound to oxidized phases of ferric oxyhydroxides and, to a lesser extent, to oxidized manganese, both of which are present in the oxidized surface layer of the sediment. In deeper sediment layers, where these metal oxides have largely been reduced, the released phosphate accumulates in the pore water and diffuses up toward the surface where it may again bind to excess FeOOH.

The distribution of the different phosphate pools in relation to Fe(III) and Mn(IV) is shown for a Danish coastal sediment in Figure 6.8 [Jensen and Thamdrup, 1993]. The dynamic pools of loosely sorbed phosphate, iron-bound phosphate, and fresh (i.e. leachable) organic phosphorus constituted nearly 50% of the total P in the uppermost 0–1 cm but were largely depleted below 3 cm depth. Only the phosphate pools bound to clay minerals ("adsorbed RP"), to calcium in apatite or calcium carbonate minerals ("apatite P"), or to refractory organic material were present below this depth. These are the main forms that become permanently buried into

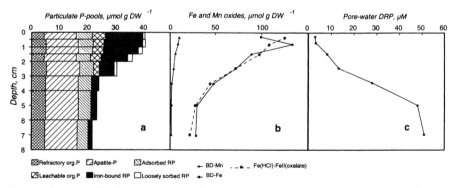

Figure 6.8. Depth distributions of phosphorus, iron and manganese in Aarhus Bay, February 1991. a: Phosphate pools analyzed by sequential extraction techniques; RP = Reactive Phosphorus. b: Iron and manganese; BD-Fe and BD-Mn = iron and manganese extracted with Bicarbonate buffered Dithionite; for comparison, Fe(III) calculated from the difference in total iron extracted with HCl and Fe(II) extracted with oxalate. c: soluble reactive phosphorus in the pore water. From Jensen and Thamdrup [1993]. Reprinted by permission of Kluwer Academic Publishers.

marine sediments [Berner et al., 1993]. Through this burial, a fraction of the total P influx to estuaries is trapped, usually only a few percent [Billen et al., 1991].

During mineralization, there is a net release of phosphate into the overlying water which is strongly regulated by the adsorption kinetics to amorphous FeOOH in the surface sediment. The $FeOOH:PO_4^{-3}$ ratio in Danish surface sediments was 6.5–16, near the equilibrium of phosphate saturation for this iron mineral phase [Jensen and Thamdrup, 1993]. Intermittent, partial reduction of the FeOOH pool due to hypoxia or sudden sedimentation of a phytoplankton bloom causes peaks in the phosphate flux into the water. The rest of the year the net flux is low and periodically may change direction toward sediment uptake [Kemp and Boynton, 1992; chapter 7]. In contrast to fresh-water sediments, in which also reduced iron minerals retain a certain capacity to bind phosphate, the Fe(II) in marine sediments mostly reacts with sulfide. As a consequence, the P retention in marine sediments is relatively less efficient than in fresh-water sediments. However, in both systems, a close coupling exists between phosphorus and iron geochemistry.

6.8. Effects of Eutrophication and Hypoxia

Among the many effects on sediments of increased organic deposition and of hypoxia in the lower water column are:

- a shift in the balance of oxidants toward anaerobic mineralization and sulfate reduction,
- reduced bottom fauna activity and a shift toward stronger diffusion limitation of solute fluxes,
- accumulation of metal sulfides and of H_2S in the pore water,
- inhibition of nitrification and preferential release of ammonium,
- release of Fe(III)-bound phosphate to the water column,
- release of H_2S in case of severe oxygen depletion,
- enhanced burial of organic carbon and nitrogen in the sediment.

The oxidants of the suboxic zone between the oxygen and sulfate respiration zones, i.e. NO_3^-, Mn(IV) and Fe(III), are more important in offshore sediments with lower organic deposition than in coastal sediments [Canfield et al., 1993]. By an increased organic flux due to eutrophication, the consumption rate of these oxidants increases more than their transport rate down into the sediment. The nitrate zone thus becomes very narrow and limiting to denitrification, manganese oxides become reduced, and Mn^{2+} is lost to the water column. Also iron oxides are reduced and the Fe(II) is bound by sulfide. Eutrophication preferentially stimulates sulfate reduction, which may even become more important than oxygen for the direct respiration of organic matter. Most of the oxygen uptake may then – directly or with Mn and Fe as intermediate electron carriers – be used for the reoxidation of the large amounts of H_2S to sulfate. The development of white films of the filamentous sulfur bacteria, *Beggiatoa* spp., on the sediment show that the H_2S is now reaching the very surface and that the buffer capacity of the intermediate oxidants has been depleted.

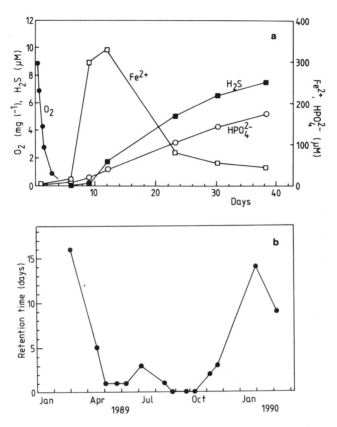

Figure 6.9. Oxic-anoxic transition of sediment from Aarhus Bay. Sediment cores were sealed gas-tight without headspace but with a 4 cm water column and incubated at in situ temperature at different times of the year. a: Chemical changes in the overlying water during 40 days incubation in fall 1989. b: time lag after depletion of oxygen before H_2S became detectable (>1 µM) in the overlying water at different times of the year. (Tina Sølbek Schmidt, unpublished data).

The transition from the well-oxidized to the reduced sediment, as a result of eutrophication and hypoxia, has been reproduced experimentally in bell jars on the sea floor [e.g., Balzer, 1984] or in sediment cores in the laboratory. The latter method has been used as an assay for the capacity of a sediment to withstand oxygen depletion before dramatic changes in chemistry and nutrient flux [T.S. Schmidt, 1990] (Figure 6.9). In late fall, the oxygen was depleted after 3–4 days, and Fe^{2+} was rapidly released from the sediment. After 9 days, the oxidation capacities of reactive Mn(IV) and Fe(III) were apparently exhausted. In addition, H_2S was released in small amounts. Two weeks after the start of the experiment, the free Fe^{2+} was gradually taken up again as it became bound by sulfide in the sediment. Concurrent with the rapid conversion of FeOOH to Fe^{2+}, phosphate was no longer bound efficiently and was released to the anoxic water. The retention time between total O_2

depletion and the first release of H_2S, which is indicative of the buffer capacity of the most reactive Mn- and Fe-oxides, was found to vary from 0 to 16 days throughout the year with the longest retention time in winter and the shortest in summer, especially during periods of hypoxia in situ (cf. Rasmussen and Jørgensen [1992]).

In conclusion, the mineralization processes in coastal sediments and the nutrient fluxes across the sediment-water interface are very sensitive to eutrophication and especially to hypoxia. The sediments function as an intermediate storage site for nitrogen and phosphorus. These nutrients are released throughout the year but mostly during the growth season of the phytoplankton. Although the burial efficiency of organic matter increases with its deposition rate [Henrichs and Reeburgh, 1987], the retention capacity of coastal sediments for N and P is limited. An important question, for which the database is still insufficient or partly contradictory, is whether the increased fluxes of N and P mainly lead to deposition in estuaries and shelf sediments, or whether N and P are exported to and buried in sediments beyond the shelf area, and whether the excess N is mostly denitrified [e.g., Walsh et al., 1985; Christensen et al., 1987].

References

Aller, R. C., The sedimentary Mn cycle in Long Island Sound: Its role as intermediate oxidant and the influence of bioturbation, O_2, and C_{org} flux on diagenetic reaction balances, *J. Mar. Res., 52*, 259–295, 1994.

Balzer, W., F. Pollehne, and H. Erlenkeuser, Cycling of organic carbon in a coastal marine system, in *Sediments and Water Interactions*, edited by P. G. Sly, pp. 323–328, Springer, New York, 1986.

Berner, R. A., Burial of organic carbon and pyrite sulfur in the modern ocean: its geochemical and environmental significance, *Am. J. Sci., 282*: 451–473, 1982.

Berner, R. A., K. C. Ruttenberg, E. D. Ingall, and J.-L. Rao, The nature of phosphorus burial in modern marine sediments, in *Interactions of C, N, P and S Biogeochemical Cycles and Global Change*, edited by R. Wollast, F. T. Mackenzie and L. Chou, pp. 365–378, NATO ASI Series, Vol. 14, Springer, Berlin, 1993.

Billen, G., C. Lancelot, and M. Meybeck, N, P, and Si retention along the aquatic continuum from land to ocean, in *Ocean Margin Processes in Global Change*, edited by R. F. C. Mantoura, J.-M. Martin, and R. Wollast, pp. 19–44, John Wiley & Sons, Chichester, 1991.

Binnerup, S. J., K. Jensen, N. P. Revsbech, M. H. Jensen, and J. Sørensen, Denitrification, dissimilatory reduction of nitrate to ammonium, and nitrification in a bioturbated estuarine sediment as measured with ^{15}N and microsensor techniques, *Appl. Envir. Microbiol., 58*, 303–313, 1992.

Blackburn, T. H., and N. D. Blackburn, Model of nitrification and denitrification in marine sediments, *FEMS Microbiol. Lett., 100*, 517–522, 1992.

Burdige, D. J., The biogeochemistry of manganese and iron reduction in marine sediments, *Earth-Sci. Rev., 35*, 249–284, 1993.

Caffrey, J. M., N. P. Sloth, H. F. Kaspar, and T. H. Blackburn, Effect of organic loading on nitrification and denitrification in a marine sediment microcosm, *FEMS Microbiol. Ecol., 12*, 159–167, 1993.

Canfield, D. E., Organic matter oxidation in marine sediments, in *Interactions of C, N, P and S Biogeochemical Cycles and Global Change*, edited by R. Wollast, F. T. Mackenzie and L. Chou, pp. 333–363, NATO ASI Series, Vol. 14, Springer, Berlin, 1993.

Canfield, D. E., B. Thamdrup, and J. W. Hansen, The anaerobic degradation of organic matter in Danish coastal sediments: iron reduction, manganese reduction, and sulfate reduction, *Geochim. Cosmochim. Acta, 57*, 3867–3883, 1993a.

Canfield, D. E., B. B. Jørgensen, H. Fossing, R. Glud, J. Gundersen, N. B. Ramsing, B. Thamdrup, J. W. Hansen, L. P. Nielsen, and P. O. J. Hall, Pathways of organic carbon oxidation in three continental margin sediments, *Mar. Geol., 113*, 27–40, 1993b.

Christensen, J. P., W. M. Smethie Jr., and A. H. Devol, Benthic nutrient regeneration and denitrification on the Washington continental shelf, *Deep-Sea Res., 34*, 1027–1047, 1987.

Dando, P. R., I. Bussmann, S. J. Niven, S. C. M. O'Hara, R. Schmaljohann, and L. J. Taylor, A methane seep area in the Skagerrak, the habitat of the pogonophore *Siboglinum poseidoni* and the bivalve mollusc *Thyasira sarsi*, *Mar. Ecol. Prog. Ser., 107*, 157–167, 1994.

Eisma, D. and J. Kalf, Dispersal, concentration and deposition of suspended matter in the North Sea, *Rapp. P.-v Réun. Cons. Perm. Intl Explor. Mer, 181*, 7–14, 1987.

Floderus, S., and L. Håkanson, Resuspension, ephemeral mud blankets and nitrogen cycling in Laholmsbukten, south east Kattegat, *Hydrobiologia, 176/177*, 61–75, 1989.

Gibbs, R. J., Sites of river derived sedimentation in the ocean, *Geology, 9*, 77–80, 1981.

Glud, R. N., J. K. Gundersen, B. B. Jørgensen, N. P. Revsbech and H. D. Schulz, Diffusive and total oxygen uptake of deep-sea sediments in the eastern South Atlantic Ocean: *in situ* and laboratory measurements, *Deep-Sea Res., 41*, 1767–1788, 1994.

Graf, G., Benthic-pelagic coupling: A benthic view, *Oceanogr. Mar. Biol. Ann. Rev., 30*, 149–190, 1992.

Graf, G., R. Schulz, R. Peinert, and L.-A. Meyer-Reil, Benthic response to sedimentation events during autumn to spring at a shallow water station in Western Kiel Bight. I. Analysis of processes on a community level, *Mar. Biol., 77*, 235–246, 1983.

Gundersen, J. K., and B. B. Jørgensen, Microstructure of diffusive boundary layers and the oxygen uptake of the sea floor, *Nature, (Lond.), 345*, 604–607, 1990.

Gundersen, J. K., R. N. Glud, and B. B. Jørgensen, Oxygen uptake of the sea floor (in Danish), *Havforskning fra Miljøstyrelsen, 57*, 155 pp., Danish Environmental Protection Agency, Copenhagen, 1994.

Hansen, L. S., and T. H. Blackburn, Effect of algal bloom deposition on sediment respiration and fluxes, *Mar. Biol., 112*, 147–152, 1992.

Hargrave, B. T., Sinking of particulate matter from the surface water of the ocean, in *Heterotrophic Activity in the Sea*, edited by J. E. Hobbie and P. J. LeB. Williams, pp. 155–178, Plenum Press, New York, 1984.

Hedges, J. I., W. A. Clark, and G. L. Cowie, Organic matter sources to the water column and surficial sediments of a marine bay, *Limnol. Oceanogr., 33*, 1116–1136, 1988.

Henrichs, S. M., and W. S. Reeburgh, Anaerobic mineralization of marine sediment organic matter: rates and the role of anaerobic processes in the oceanic carbon economy, *Geomicrobiol. J., 5*, 191–237, 1987.

Hüttel, M., and G. Gust, Impact of bioroughness on the interfacial solute exchange in permeable sediments, *Mar. Ecol. Prog. Ser., 89*, 253–267, 1992.

Ingall, E. D, and P. van Cappellen, Relation between sedimentation rate and burial of organic phosphorus and organic carbon in marine sediments. *Geochim. Cosmochim. Acta, 54*, 373–386, 1990.

Jensen, H. S., and B. Thamdrup, Iron-bound phosphorus in marine sediments as measured by bicarbonate-dithionate extraction, *Hydrobiologia, 253*, 47–59, 1993.

Jensen, P., I. Aagaard, R. A. Burke, P. R. Dando, N. O. Jørgensen, A. Kuijpers, T. Laier, S. C. M. O'Hara, and R. Schmaljohann, "Bubbling reefs" in the Kattegat: submarine landscapes of carbonate-cemented rocks support a diverse ecosystem at methane seeps, *Mar. Ecol. Prog. Ser., 83*, 103–112, 1992.

Jørgensen, B. B., and N. P. Revsbech, Colorless sulfur bacteria, *Beggiatoa* spp. and *Thiovulum* spp., in O_2 and H_2S micrograudients, *Appl. Envir. Microbiol., 45*, 1261–1270, 1983.

Jørgensen, B. B., and N. P. Revsbech, Diffusive boundary layers and the oxygen uptake of sediments and detritus, *Limnol. Oceanogr., 30*, 111–122, 1985.

Jørgensen, B. B., and N. P. Revsbech, Oxygen uptake, bacterial distribution, and carbon-nitrogen-sulfur cycling in sediments from the Baltic Sea – North Sea transition, *Ophelia, 31*, 29–49, 1989.

Jørgensen, B. B., M. Bang, and T. H. Blackburn, Anaerobic mineralization in marine sediments from the Baltic Sea – North Sea transition, *Mar. Ecol. Prog. Ser., 59*, 39–54, 1990.

Kemp, W. M., and W. R. Boynton, Benthic pelagic interactions: nutrient and oxygen dynamics, in *Oxygen Dynamics in the Chesapeake Bay*, edited by D. E. Smith, M. Leffler, and G. Mackiernan, pp. 149–221, Maryland Sea Grant, College Park, Maryland, 1992.

Kemp, W. M., P. Sampou, J. Caffrey, M. Mayer, K. Henriksen, and W. R. Boynton, Ammonium recycling versus denitrification in Chesapeake Bay sediments, *Limnol. Oceanogr., 35*, 1545–1563, 1990.

Klump, J. V., and C. S. Martens, Biogeochemical cycling in an organic-rich coastal marine basin. 5. Sedimentary nitrogen and phosphorus budgets based upon kinetic models, mass balances, and the stoichiometry of nutrient regeneration, *Geochim. Cosmochim. Acta, 51*, 1161–1173, 1987.

Larsen, B., Geography of the sea (in Danish), in *Danmarks Natur*, Vol. 3, edited by A. Nørrevang and T. J. Meyer, pp. 9–23, Politiken, Copenhagen, 1968.

Lomstein, B. Aa., and T. H. Blackburn, The nitrogen cycle of the sea floor in Aarhus Bay (in Danish), *Havforskning fra Miljøstyrelsen, 16*, 74 pp., Danish Environmental Protection Agency, Copenhagen, 1992.

Lomstein, B. Aa., T. H. Blackburn, and K. Henriksen, Aspects of nitrogen and carbon cycling in the northern Bering Shelf sediment. I. The significance of urea turnover in the mineralization of NH_4^+. *Mar. Ecol. Prog. Ser., 57*, 237–247, 1989.

Lovley, D. R., Organic matter mineralization with the reduction of ferric iron, A review, *Geomicrob. J., 5*, 375–399, 1987.

Lovley, D. R., and E. J. P. Phillips, Competitive mechanisms for inhibition of sulfate reduction and methane production in the zone of ferric iron reduction in sediments, *Appl. Envir. Microbiol., 53*, 2636–2641, 1987.

Madsen, P. F., and B. Larsen, Accumulation of mud sediments and trace metals in the Kattegat and the Belt Sea, 68 pp., *Rep. Marine Pollution Lab., 10*, Charlottenlund, Denmark, 1986.

Mantoura, R. F. C., J.-M. Martin, and R. Wollast (Eds), *Ocean Margin Processes in Global Change*, 469 pp., John Wiley & Sons, Chichester, 1991.

Meyer-Reil, L.-A., Benthic response to sedimentation events during autumn to spring at a shallow station in the Western Kiel Bight. II. Analysis of benthic bacterial populations, *Mar. Biol., 77*, 247–256, 1983.

Nielsen, L. P., P. B. Christensen, and S. Rysgaard, Denitrification in fjords and coastal waters (in Danish), *Havforskning fra Miljøstyrelsen, 50*, 52 pp., Danish Environmental Protection Agency, Copenhagen, 1994.

Nixon, S. W., Remineralization and nutrient cycling in coastal marine ecosystems, in *Nutrients and Estuaries*, edited by B. J. Neilson and L. E. Cronin, pp. 111–138, Humana Press, New York, 1981.

Olesen, M., and C. Lundsgaard, Seasonal sedimentation of autochthonous material from the euphotic zone of a coastal system, *Estuar. Coast. Shelf Sci.*, 1995, in press.

Pedersen, H., B. Aa. Lomstein, and T. H. Blackburn, Evidence for bacterial urea production in marine sediments, *FEMS Microbiol. Ecol.*, *12*, 51–59, 1993.

Pejrup, M, J. Valeur, and A. Jensen, Vertical sediment fluxes in the Aarhus Bay, Denmark, measured by use of sediment traps, *Cont. Shelf. Res.*, submitted.

Rasmussen, H., and B. B. Jørgensen, Microelectrode studies of seasonal oxygen uptake in a coastal sediment: role of molecular diffusion, *Mar. Ecol. Prog. Ser.*, *81*, 289–303, 1992.

Revsbech, N. P., B. B. Jørgensen, and T. H. Blackburn, Oxygen in the sea bottom measured with a microelectrode, *Science, 207*, 1355–1356, 1980.

Seitzinger, S. P., Denitrification in freshwater and coastal marine ecosystems: ecological and geochemical significance, *Limnol. Oceanogr.*, *33*, 702–724, 1988.

Sørensen, J., B. B. Jørgensen, and N. P. Revsbech, A comparison of oxygen, nitrate and sulfate respiration in coastal marine sediments, *Microbiol. Ecol.*, *5*, 105–115, 1979.

Smetacek, V., Annual cycle of sedimentation in relation to plankton ecology in western Kiel Bight, *Ophelia*, Supplement *1*, 65–76, 1980.

Smetacek, V., Role of sinking in diatom life history cycles: ecological, evolutionary and geological significance, *Mar. Biol.*, *84*, 239–251, 1985.

Suess, E., Particulate organic carbon flux in the oceans: surface productivity and oxygen utilization, *Nature, 288*, 260–263, 1980.

Thamdrup, B., R. N. Glud, and J. W. Hansen, Manganese oxidation and *in situ* manganese fluxes from a coastal sediment, *Geochim. Cosmochim. Acta, 58*, 2563–2570, 1994.

Valeur, J. R., M. Pejrup, and A. Jensen, Particulate nutrient fluxes in Vejle Fjord and Aarhus Bay (in Danish), *Havforskning fra Miljøstyrelsen, 14*, 127 pp., Danish Environmental Protection Agency, Copenhagen, 1992.

van Weering, T. C. E., G. W. Berger, and J. Kalf, Recent sediment accumulation in the Skagerrak, northeastern North Sea, *Neth. J. Sea Res., 21*, 177–189, 1987.

Walsh, J. J., E. T. Premuzic, J. S. Gaffney, G. T. Rowe, G. Harbottle, R. W. Stoenner, W. L. Basalm, P. R. Betzer, and S. A. Mackoll, Organic storage of CO_2 on the continental slope off the mid-Atlantic bight, the southeastern Bering Sea, and the Peru coast, *Deep-Sea Res., 32*, 853–883, 1985.

Wollast, R., The coastal organic carbon cycle: fluxes, sources, and sinks, in *Ocean Margin Processes in Global Change*, edited by R. F. C. Mantoura, J.-M. Martin, and R. Wollast, pp. 365–381, John Wiley & Sons, Chichester, 1991.

7

Case Study – Aarhus Bay

Bo Barker Jørgensen

7.1. Introduction

Aarhus Bay is a hydrographically open coastal ecosystem situated at the Baltic Sea-Kattegat transition. An intensive field study of the biological, chemical and physical processes in the water column and in the sediments was carried out in 1990–1991 with the main goal of understanding the seasonal element cycling and the nutrient coupling between water and sediment. Specific questions were:

– how much of the primary production sinks out and is mineralized in the sediment?
– how does nutrient release from the sediment affect primary production?
– how do eutrophication and hypoxia affect the chemical environment and biological processes in the sediment?

The emphasis of this study was on direct measurements of fluxes and process rates. A single sampling station, "Station 6", at 16 m water depth in the central bay was studied for 1.5 years with participation from the Danish Universities of Aarhus, Copenhagen and Odense, the County of Aarhus, as well as the Hydraulic Institute, the Water Quality Institute, and the National Environmental Research Institute. Through the chosen approach of the study, a detailed coherent data set was obtained for one location, based on which an annual budget of the main element cycles could be calculated. The transports over the boundaries of the open bay system were, however, poorly defined and the budget could therefore not be fully balanced.

7.2. Aarhus Bay

Aarhus Bay covers an area of 320 km² and has a mean water depth of 15 m. It deepens in the eastern part and connects to the southern Kattegat and the Belt Sea over shallows and through narrow troughs (Figure 7.1). Geologically, it was formed during the last glaciaton by ice margin oscillations which left terminal

Eutrophication in Coastal Marine Ecosystems
Coastal and Estuarine Studies, Volume 52, Pages 137–154
Copyright 1996 by the American Geophysical Union

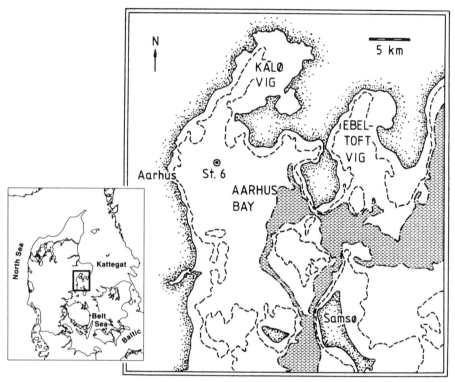

Figure 7.1. Aarhus Bay and the position of the sampling station, Station 6 (56°09''10' N, 10°19''20' E). Depth curves: – – – 10 m, – · – 20 m; water depths >20 m are shaded. Inset: Map of the Baltic Sea - North Sea transition showing location of Aarhus Bay.

moraines and dead-ice formations. On top of quaternary till and glaciafluvial deposits, the sedimentation basin of Aarhus Bay is today gradually filling up at a mean rate of ca. 2 mm yr^{-1} (Pejrup et al., submitted). A determination of the total thickness of deposited postglacial mud is, however, difficult due to seismic scattering by methane bubbles which are present below the sulfate zone, from 3–4 m sediment depth. The sediment consists of 21% sand, 23% silt and 56% clay and has an organic content of 2.1–3.7% org. C [Valeur et al., 1992].

7.3. Hydrography and Plankton

During the field study, the water column was stratified during 70–80% of the year with Baltic surface water of 15–25 psu overlying saline Kattegat water of 29–33 psu [Sørensen and Nielsen, 1992] (Figure 7.2). During spring and early summer, this stratification is further stabilized by surface water heating, and the salinity and temperature *difference* between top and bottom may reach 13 psu and 11 °C, respectively. The exchange of water masses follows closely hydrographic events in the south-

Depth, m

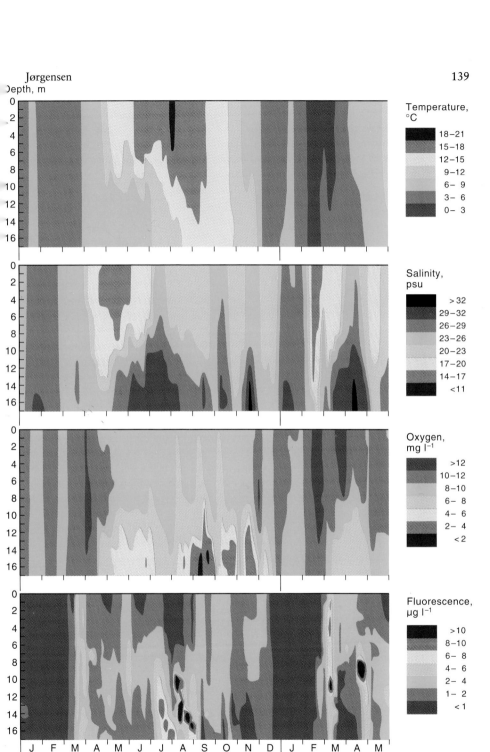

Figure 7.2. Water column distribution of, a) temperature, b) salinity, c) oxygen, and d) induced fluorescence during the period, January 1990–May 1991. Data from Sørensen and Nielsen [1992].

ern Kattegat, which tilt the pycnocline and force deep Kattegat water into or out of the bay according to the regional wind field. The calculated mean residence time of water in the bay is only 12 days and the transport distance during one current event is usually short, 2–5 km, after which the current reverses [Christiansen et al., 1994]. Tidal currents are less important, as the tidal amplitude is only 20–40 cm.

The lower water column became gradually depleted in oxygen as stratification became stronger during spring, due to the degradation of plankton and detritus sinking down from the euphotic zone (Figure 7.2c). The oxygen concentration at air saturation of the bottom water corresponded to about 270 μM in summer and 350 μM in winter. Hypoxia started in April and March, respectively, of the two years studied, shortly after the collapse of a spring phytoplankton bloom. More extensive oxygen depletion, down to < 2 mg O_2 l^{-1} (i.e. < 50 μM O_2 or 15–20% of air saturation), occurred during late summer and fall, with lowest oxygen concentrations during periods when the zooplankton grazing pressure was relatively low and a larger fraction of the phytoplankton was therefore degraded at the sediment surface [Sørensen and Nielsen, 1992]. Deeper water from southern Kattegat flowed into the bay on several occasions and, as oxygen was gradually depleted from this partly isolated water mass, resulted in the sudden death of many demersal fish and benthic invertebrates. This strong, intermittent hypoxia is apparent in Figure 7.2, which shows the temporal coincidence of low oxygen and high salinity of the bottom water. Measurements carried out with a benthic microsensor instrument or with a special bottom water sampler revealed vertical oxygen gradients also in the lowest few meters or decimeters over the sediment on occasions when a stable pycnocline just over the sea bed prevented mixing of the bottom water [Gundersen et al., 1994; cf. Jørgensen, 1980]. On one occasion oxygen dropped from 280 to 65 μM through the bottom 2 m of water (chapter 6, Figure 6.6, left). This small-scale hypoxia was less significant for the general oxygen balance, but was critical to the benthic invertebrates, which were thereby exposed to even lower oxygen levels than indicated at the depth resolution of conventional hydrographic instruments (Figure 7.2).

The spring bloom of diatoms, mostly *Thalassiosira* spp., which over a few weeks assimilated the nutrients accumulated during winter, occurred in the period of late February to early April. As is also the case in the Kattegat [Richardson and Christoffersen, 1991], the spring bloom constituted only a small fraction of the annual production, 14% in 1990, but with high photosynthesis rates, up to 180 mmol C m^{-2} d^{-1} (= 2.2 g C m^{-2} d^{-1}), over a short period (Kruse, submitted) Figure 7.3. The diatoms sank out of the water column within a few days after the nutrient depletion, and apparently intact diatom cells with a high chlorophyll content and a C:N ratio of 6–7, similar to the Redfield ratio for living plankton, appeared as a flocculent, orange detritus film on the sediment. As is typical of many eutrophic coastal systems [Billen et al., 1991], silica was depleted before nitrogen and phosphorus and the phytoplankton community changed to flagellated forms, which were mainly P-limited during spring and N-limited later in the summer [Sørensen and Nielsen, 1992]. The algae were strongly grazed during summer and detritus settled as fecal pellets. A bloom of dinoflagellates, *Ceratium* spp., developed in the fall just below the pycnocline (Figure 7.2d), where the algae utilized

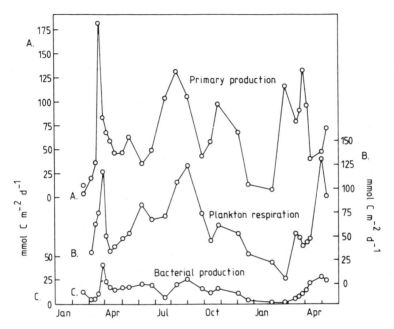

Figure 7.3. Biological activity in the water column during 1990–1991 calculated per square meter, based on measurements in incubated water samples from ten depths. Primary production was measured by the ^{14}C ABM-method. Plankton respiration was measured from the decrease in oxygen over several hours in water samples incubated in gas-tight plastic bags. Bacterial production was measured by the ^{14}C-thymidine incorporation method. Data from B. Kruse (submitted).

nutrients continuously released from the sediment. Such pycnocline blooms also develop in the Kattegat outside the bay [Nielsen et al., 1990; Bjørnsen et al., 1993]. The large, spiny *Ceratium* cells were not extensively grazed and reached the bottom as whole cells.

Respiration in the water column mostly followed variations in the seasonal plankton productivity, though often with a time lag (Figure 7.3). A comparison between photosynthesis and total respiration shows which fraction of the phytoplankton production was mineralized through the grazing food chain and microbial loop, and how much settled out of the water column. During the whole year, 68% of the fixed carbon was respired in the water, and 32% may thus have reached the sediment, with the relatively highest contributions coming from the rapidly settling diatom and *Ceratium* blooms. This percentage of sedimentation and benthic remineralization falls in the central range of values published for shallow coastal ecosystems [e.g., Suess, 1980; Nixon, 1981]. The bacterial production reflected the respiration through the year with small maxima following blooms of phytoplankton. Assuming that the bacteria allocate half the assimilated organic carbon to the measured cell growth and use half for respiration, the total carbon flow through the microbial loop would be 48% of the measured primary production (B. Kruse, in prep.).

7.4. Sedimentation and Resuspension

Vertical sediment fluxes were measured by biweekly harvesting of sediment traps permanently situated at nine depths in the water column [Valeur et al., 1995] (Pejrup et al., submitted). Due to the presence of a pycnocline, which prevented vertical mixing during 70–80% of the year, it was possible to distinguish between the net flux of primary sediment from above the pycnocline and the much larger flux of resuspended material from below (cf. chapter 6, Figure 6.2). Only the settling of the *Ceratium* bloom, which developed just below the pycnocline, could not be distinguished by this approach.

Currents and waves caused resuspension of sediment in the relatively shallow Aarhus Bay, but the erodibility varied strongly throughout the year [Lund-Hansen et al., 1993; Valeur et al., 1995]. A fine detritus layer, deposited by the spring bloom was easily brought into suspension. During fall and winter, repeated resuspension events removed the flocculent material and left a more consolidated sediment surface which largely resisted resuspension, even at measured current velocities of up to 28 cm s^{-1}. There was a positive correlation between bottom shear stress, i.e. current velocity, and resuspension measured at a 1-day time resolution (Floderus and Lund-Hansen, submitted). The resuspension was, however, apparently also affected by wave pumping which may cause a swelling and loosening, "liquefaction", of the surface sediment and bring it into suspension; cf. Maa and Metha [1987]. During the strongest storms, the amount of material in suspension >0.5 m above the bottom corresponded to a mean sediment layer of 2 mm cycling in the water at one time. The depth of erosion is, however, likely to be variable, dependent on sediment topography etc., and may locally be much deeper than a few mm.

These phenomena were demonstrated by an extreme resuspension event shown in Figure 7.4. Brackish Baltic water initially overlayed the sea floor, but after storms on December 26 and 27, saline bottom water from the Kattegat flowed into Aarhus Bay. Due to the close bottom contact of this heavy water, sediment material was entrained, and it thus carried a higher suspension load (ca. 10 mg d.wt. l^{-1}) than did the brackish water (1–2 mg l^{-1}). This is also seen from the higher turbidity (Figure 7.4). Sedimentation rates measured at daily intervals 1 m above the bottom showed resuspension peaks which did not coincide with maximum current velocities but rather with maximum wave heights and orbital velocities of the bottom water (although the highest orbital velocity reached on 26 December (6 cm s^{-1}) was only one third of the highest current velocity). On that occasion, the sediment trap at 1.5 m above the bottom collected 500 g m^{-2} of sediment material within 24 h (Floderus and Lund-Hansen, submitted). Only five times during the year were orbital velocities of >5 cm s^{-1} reached, with the maximum being 12 cm s^{-1} [Floderus, 1993].

The cycle of resuspension-resedimentation is important for the coupling of processes at the sediment-water interface. High respiration rates, rapid oxygenation of sediment iron sulfides, rapid release of nutrients, etc. may occur during a storm such as that illustrated in Figure 7.4. However, field sampling and flux measure-

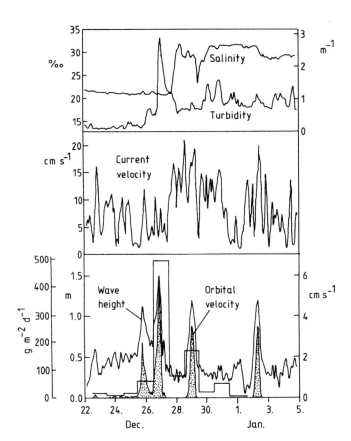

Figure 7.4. Physical mechanisms causing resuspension during a windy period in winter 1990–1991. Upper frames: salinity, turbidity, and current velocity at 0.5 m above the bottom. Bottom frame: wave heights and induced orbital velocities over the bottom computed from local wind data. The velocity of the oscillating water movement, rather than the total current velocity, correlated closely to the total sediment flux (net deposition plus resuspension) per day measured at 1 m above bottom (shown as horizontal lines). The highest value, measured on December 26–27, corresponded to 500 g d.wt per m². Data from Valeur et al. [1992] and Floderus and Lund-Hansen (submitted).

ments are difficult (and therefore not available) for such occasions, which therefore causes an underestimation of certain fluxes. It was calculated that a sediment particle was typically resuspended at least 60 times to >1 m above bottom before it was trapped below the 2.5-cm deep sediment layer of highest mixing rate (Pejrup et al., submitted). The mean settling velocity of resuspended particles was 0.5–2 m h⁻¹ during the year. Consequently, the resuspended material had a short residence time in the water, and the resuspension followed changes in current and wave energy within hours. As a result, the lateral transport distance of sediment material during each resuspension event was only a small fraction of the typical water transport distance of 2–5 km [Christiansen et al., 1994]. Furthermore, the measured resuspension decreased exponentially with height above the sediment, and over the year the material trapped at 4 m above bottom was less than 10% of that trapped 0.5 m above (cf. chapter 6; Figure 6.2). The close balance between the net deposition and the accumulation of sediment suggested that most of the primary sediment remained trapped within the sedimentation basin of Aarhus Bay. Thus, the total net deposition measured by the sediment trap method was 990 g d.wt. m^{-2} yr^{-1}, while the net accumulation of sediment calculated from ^{210}Pb profiles was 890 g d.wt. m^{-2} yr^{-1}.

7.5. Sediment Oxygen Uptake

The aerobic mineralization of organic material in the sediment was measured both in the field and in laboratory-incubated cores as the total and the diffusive uptake of oxygen (Figure 7.5). Seasonal in situ data on diffusive oxygen uptake, obtained by a benthic microelectrode instrument, "Profilur" [Gundersen and Jørgensen, 1990], showed how uptake rates depended on a combination of fresh detritus influx and of the oxygen concentration in the overlying water (Gundersen et al., in prep.). Over the year, oxygen consumption showed only little dependence on the temperature which varied between 0 and 17°C. A spring peak in oxygen uptake was followed by a period of lower respiration rate during summer while oxygen in the bottom water was low. An intermittent inflow of more oxic water in late September immediately enhanced the oxygen uptake, which indicated that the microbial respiration was limited by oxygen availability during summer. Lower rates persisted throughout winter 1990–1991 until a phytoplankton bloom sedimented the next spring. The thickness of the oxic zone, which varied between a minimum of 0.6 mm in summer and a maximum of 4 mm in winter, also closely reflected the combined effect of oxygen concentration in the water and the potential sediment O_2 consumption rate.

The total oxygen uptake measured in incubated cores was only 20% higher than the calculated one-dimensional diffusive uptake. This apparently indicating that biopumping, current-induced advective flow or surface topography played little role [Gundersen et al., 1994]. The core incubation, however, underestimates the effect of large infauna, which is not well represented or is less active in the cores. Concurrent studies on the sea floor with a microsensor and a flux chamber instrument showed that the total oxygen uptake under the least disturbed in situ conditions was 60–70% higher than the diffusive uptake.

The core incubation also does not include the effect of resuspension of the surface layer which occurs frequently in situ (see chapter 6). Experimental resuspension in the laboratory of sediment from just below the oxygen zone showed an extremely high initial oxygen consumption rate, ca. 100-fold higher than the specific rates measured in situ [Hansen et al., 1994]. Especially the rapid oxidation of pyrite, FeS_2, and other reduced iron minerals was important for the enhanced oxygen consumption, the rate of which decreased two orders of magnitude within a few hours. By combining these experimental results with field data on actual resuspension events, the potential oxygen depletion in the lower meters of the water column during resuspension could be calculated. The results indicated that the initial, rapid ef-

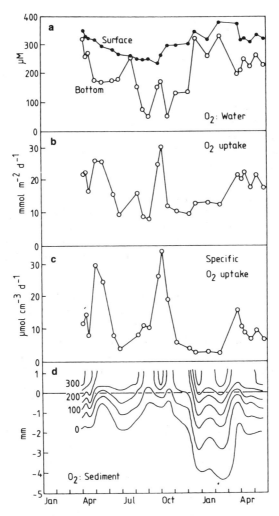

Figure 7.5. Oxygen distribution and consumption rates during 1990–1991. a) O₂ concentrations in surface water and 10 cm over the bottom. a) Diffusive O₂ uptake rates of the sediment calculated from the in situ O₂ gradient in the diffusive boundary layer multiplied with the diffusion coefficient of O₂ at the prevailing temperature and salinity. c) Mean volume-specific O₂ consumption rates calculated from the diffusive O₂ uptake of the sediment divided by the depth of the oxic zone. d) Isopleths of O₂ distribution in the sediment; numbers on curves indicate O₂ concentrations in μM. Data from Gundersen et al. [1994].

fect was small, only about 3% of the oxygen pool at air saturation in a 1 m water column would be consumed within an hour by resuspension of the top 5 mm of sediment. Local effects on oxygen at the sediment-water interface may, however, be more significant, and the resuspension converts iron and manganese minerals to an oxidized state which is more reactive to sulfide.

7.6. Anaerobic Mineralization

Below the oxic zone, most oxidation of organic matter was due to sulfate reduction. Sulfate became depleted at 2.5 m depth, below which methane accumulated and formed gas bubbles below ca. 3.5 m. The rate of sulfate reduction largely followed the seasonal temperature variation, and the zone of maximum reduction rate moved upward during summer, from 5–6 cm to 2–3 cm, as the sediment became more reduced and sulfidic [Thamdrup et al., 1994; Figure 7.6]. A distinct zone of high sulfate reduction rate within the top 5 mm suddenly appeared with a delay of a few weeks after precipitation of the spring diatom bloom (data not shown). In some years, this zone was visible for several weeks as a thin, black surface band of iron sulfide overlying the brownish oxidized sediment zone which, during early spring, extended to several cm depth [Moeslund et al., 1994]. The microbial degradation of the freshly settled phytodetritus was thus so intensive that an anoxic, sulfidic microzone developed on top of the sediment. Deeper in the sediment, sulfate reduction proceeded at rapidly decreasing rates, down to the depth of complete sulfate depletion at 2.5 m.

Mass balance calculations showed that 85% of the H_2S formed from sulfate reduction was again reoxidized through the redox cascade of oxidized iron and manganese, nitrate and oxygen (chapter 6). Especially the top few cm of sediment were rich in reactive iron and manganese oxides, with ca. 10-fold more iron than manganese [Thamdrup et al., 1994; Figure 7.7]. Below 2–3 cm depth, the concentration of extracted oxidized iron dropped to a background level of non-reactive Fe(III) of ca. 20 μmol cm^{-3}. The calculated turnover time of the total manganese pool was 1–3 weeks and of the reactive iron pool 2–6 months. The main zones of metal reduction were evident from maxima of dissolved Fe^{2+} and Mn^{2+}, which occurred at 2–3 cm and at 1 cm depth, respectively. A high Mn^{2+} peak was observed in April as the result of intensive manganese reduction just below the freshly settled phytoplankton detritus. Recycling of the reduced metals took place through upward diffusion along steep surface gradients and subsequent oxidation with oxygen.

Free H_2S was detected only below the zone of reactive iron (Figure 7.7). With a calculated turnover time of <10 min, the large amounts of H_2S formed in the main sulfate reduction zone must thus have been rapidly reoxidized or precipitated by the metal oxides, which were transported downward into the sediment by bioturbation. Quantitative evidence for anaerobic sulfide reoxidation through down-mixing of metal oxides has been lacking, but in the Aarhus Bay sediment this mechanism was indeed confirmed by calculations of mass balance and transport coefficients.

The maintenance of a zone of oxidized metals through continuous Mn^{2+} and Fe^{2+} oxidation by O_2 is important to prevent the emission of H_2S into the overlying water. Laboratory experiments showed that the total MnO_2 pool could be depleted within one month in the absence of oxygen. Accordingly, the extensive hypoxia during summer 1990 created total MnO_2 depletion and low iron oxide levels for 3–4 weeks in October (Figure 7.7). A large Mn^{2+} pool was released to the overlying

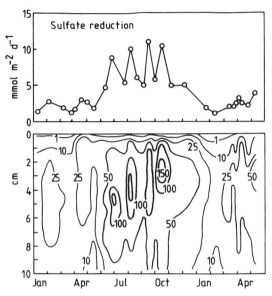

Figure 7.6. Sulfate reduction in the top 0–10 cm of sediment during 1990–1991. Top: Integrated rates per m^2 of sediment. Bottom: Isopleths of reduction rates (nmol SO_4^{2-} cm^{-3} d^{-1}). Data from Thamdrup et al. [1994]. Reprinted with permission from Elsevier Science Ltd.

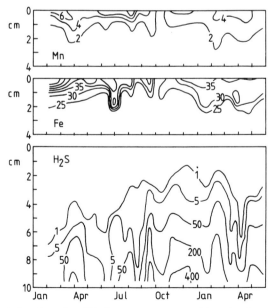

Figure 7.7. Isopleths of total, solid-phase manganese and iron in the sediment and of H_2S in the pore water during 1990–1991. Concentrations on curves are in μmol cm^{-3} for the metals and in μM for H_2S. Data from Fossing et al. [1992].

water, but soon returned as the Mn^{2+} was oxidized to MnO_2 higher up in the water column and settled again, trapped together with other sedimenting particles [Thamdrup et al., 1994].

7.7. Nutrient Cycling

The exchangeable phosphate in the sediment occurred mostly adsorbed to iron hydroxides and was partly released during late summer as the iron gradually became reduced and reacted with sulfide [Jensen and Thamdrup, 1993; Jensen et al., 1995]. During the initial degradation of the spring plankton bloom on the sediment surface, a short pulse of dissolved phosphorus release from the organically bound phosphate also occurred (Figure 7.8). As the manganese and iron oxide pools became depleted in fall 1990, a much larger pulse of phosphorus was emitted from the sediment because the FeOOH-adsorbed phosphate could no longer be retained. This caused a transient accumulation of phosphate in the bottom water of up to 2.0 and 3.4 μM HPO_4^{2-} in 1990 and 1991, respectively [Sørensen and Nielsen, 1992]. Some of the phosphate was trapped again by adsorption to the metal oxides in the water column and subsequently sedimented. The short period of this open "iron-manganese window" was thus important for the nutrient cycling in the sediment,

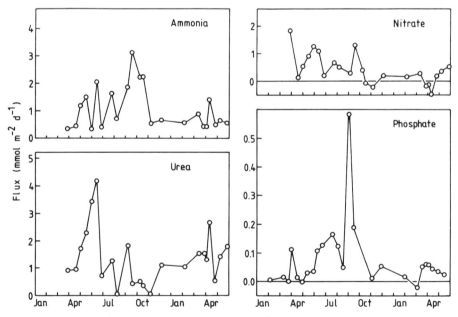

Figure 7.8. Nutrient fluxes across the sediment-water interface measured in laboratory incubated cores during 1990–1991. Cores were kept at in situ temperature under in situ water until incubation on the day of sampling. The P-flux includes the total dissolved P. Positive values indicate flux from sediment to water. Data from Lomstein and Blackburn [1992] and Jensen et al. [1995].

but rapid recycling via the water column largely retained the elements within the bay. The large phosphorus pulse occurred in the fall when low nitrogen availability and fading daylight limited the primary production. Accordingly, it had no immediate effect on the plankton productivity. Experimental studies showed that additional release of phosphate may have taken place during resuspension events and may thus account for a discrepancy between measured and calculated release (H.S. Jensen, pers. comm.).

There was a preferential loss of phosphate from the fresh plankton detritus as it settled out of the water column, and the C:P ratio increased from a Redfield ratio of 106 in the algae to 155 in those sediment traps with the least resuspended material. Phosphate, however, became trapped in the sediment and buried more efficiently than organic carbon. Below the mixing zone in the sediment the C:P ratio was ca. 133.

An opposite trend was found for organic nitrogen, which was depleted in the sediment relatively faster than organic carbon. Although the nitrogen was also preferentially lost from the sedimenting detritus, whereby the C:N ratio increased from the Redfield ratio of 6.5 up to 8.1 in the sediment traps, the C:N ratio below the mixing zone of the sediment was ca. 11 (Lomstein and Blackburn, submitted; chapter 6, Figure 6.7). The nitrogen was mainly buried as organic N and only 5% as adsorbed or free NH_4^+. Compared to the 21% of the deposited organic carbon, which was buried into deeper sediment layers, the burial of deposited nitrogen was only 15%, whereas of phosphorus it was 32% (Figure 7.9). Nitrogen is thus preferentially released during mineralization, while the sediment tends to trap phosphate. Furthermore, a reactive pool of phosphate is accumulated in the oxidized surface sediment, bound to reactive Fe(III) minerals, and may send pulses of phosphate up into the water column during hypoxia. This phosphate pool remains at the surface and is not buried because the phosphate-binding Fe(III) minerals become reduced deeper in the sediment and thus release the phosphate again to diffuse upward.

While 15% of the deposited particulate organic nitrogen was buried in the sediment, the rest was degraded and converted into dissolved organic nitrogen (DON), urea, and inorganic nitrogen [Lomstein and Blackburn, 1992]. Radiotracer studies of ^{14}C-urea turnover showed its major importance as a product in organic degradation and as an intermediate in the formation of NH_4^+; cf. Lund and Blackburn [1989]. Nitrogen was thus released from the sediment to the water at relatively constant rates, mostly in the form of urea and NH_4^+ (Figure 7.8). The former dominated in the spring when the sediment was rather oxidized and the urea concentration in the overlying water was low. During periods of oxygen depletion in late summer, the urea was further degraded in the sediment to NH_4^+, which was then released.

The seasonal variation of nitrate flux from the sediment is regulated by several factors including nitrification, denitrification, nitrate concentration in the overlying water, and temperature [e.g., Kemp et al., 1990]. Nitrification was low during winter and in periods when the oxic zone was thin, < 1 mm, i.e. after sedimentation of the spring phytoplankton bloom and during summer [Jensen et al., 1990]. The mean diffusion time through a 1 mm deep O_2 zone is only ca. 10 min, and such a thin oxic zone combined with the short retention time may hinder the efficient ammonium uptake by the nitrifying bacteria. The nitrification of an estimated 40% of the NH_4^+ consumed 0.5 mol O_2 yr^{-1} or 5–8% of the annual sediment oxygen uptake. During winter, the nitrate concentration of the bottom water increased to near 20 µM and, in periods, the sediment was taking up nitrate (Figure 7.8). Over a whole year, the total budget of deposited particulate organic nitrogen (PON) was: 15% buried as PON, 12% released as DON, 30% released as urea, 26% released as NH_4^+, and 18% nitrified to NO_3^- of which half (9%) was released as NO_3^- and the other 9% denitrified to N_2 [Valeur et al., 1992; Lomstein and Blackburn, 1992].

7.8. Budget and Conclusions

In the calculation of an annual budget for processes of the main element cycles, as is done in the following (Figure 7.9), much important information is lost on the dynamics and regulation of the processes which can be read only out of the seasonal data. The annual balance is, however, useful in the understanding of the coupling of processes in the water column and sediment and to check the internal consistency of the seasonal data. A simplified scheme is presented here, in which data have been calculated for a 1-m^2 area of the sea floor over a year. The production or consumption of oxygen and other oxidants is given in molar units to facilitate com-

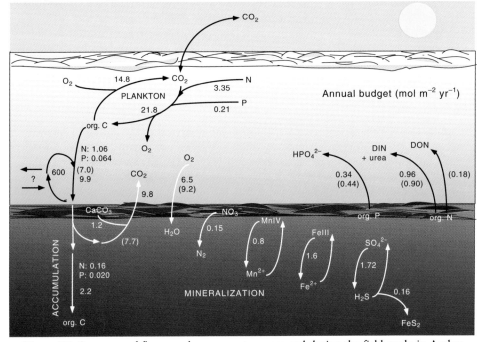

Figure 7.9. Summary of fluxes and process rates measured during the field study in Aarhus Bay, May 1990–May 1991. Numbers in parentheses were derived by difference while the rest are based on independent rate measurements and calculations. Units are given in mol m^{-2} yr^{-1} for each element.

parison to the carbon transformations and element cycling by taking the reaction stoichiometry of the elements, C, O, N, Mn, Fe, and S into account (cf. chapter 6, Table 6.1). It should be noted that almost all data are based on direct process measurements and that a balanced budget is, therefore, not to be expected, mainly because the system is open and with a poorly defined lateral exchange, but also because of possible inaccuracies in the measured rates.

The annual net photosynthesis of phytoplankton was 21.8 mol C m^{-2} yr^{-1} (260 g C m^{-2} yr^{-1}), which places Aarhus Bay in the upper mesotrophic range of coastal ecosystems. Given a Redfield ratio of elemental composition, the concurrent assimilation of nitrogen and phosphorus nutrients was 3.35 mol N and 0.21 mol P. The measured planktonic oxygen respiration was equivalent to 14.8 mol C or 68% of the primary production, and the remaining 7.0 mol C or 32% may thus have been deposited on the sea floor (B. Kruse, submitted). Sediment trap studies concluded that a somewhat larger amount of detritus, 9.9 mol C equivalent to 45%, had settled from the photic zone [Valeur et al., 1992].

Due to frequent resuspension of the surface sediment, combined with efficient mixing of the upper 2–3 cm as shown by ^{210}Pb data [Valeur et al., 1992], deposited detritus was recycled through the lower water column many times before burial into deeper layers. The measured gross vertical flux of organic carbon at 1.5 m above the sea floor was 600 mol C, i.e. 60-fold higher than the net deposition. The resuspension cycle was even more intensive closer to the sediment surface (cf. chapter 6, Figure 6.2).

Of the deposited organic carbon, 2.1 mol was buried below the bioturbated layer while the remaining 7.8 mol was mineralized to CO_2. Another 1.2 mol CO_2 was formed by dissolution of 93% of deposited carbonate shells [Valeur et al., 1992]. The measured CO_2 release was 9.8 mol, corresponding well to the net carbon deposition. The total oxygen uptake measured in sediment cores, 6.5 mol O_2, was too low to account for the oxidized organic carbon, but it is also likely to be an underestimate, as discussed above. The calculated total uptake based on a comparison with in situ flux chamber measurements was 9.2 mol O_2 [Gundersen et al., 1994].

Of the oxidants in the suboxic zone (cf. chapter 6), NO_3^- could account for only $0.15 \times 5/4 = 0.18$ mol C oxidized, and denitrification was thus insignificant for the carbon oxidation in the sediment (Lomstein and Blackburn, submitted). Manganese and iron reductions were apparently involved in the oxidation and precipitation of sulfide rather than directly in the oxidation of organic carbon. With a redox stoichiometry of 2 mol Mn(IV) or 4 mol Fe(III) to oxidize 1 mol org. C, they were also of minor significance as indirect carbon oxidants, each equivalent to ca. 0.4 mol org. C [Thamdrup et al., 1994].

Sulfate reduction was 1.72 mol and could account for 3.4 mol org. C or 44% of the organic carbon oxidized in the sediment [Fossing et al., 1992]. Of the H_2S formed, 0.25 mol or 15% was buried mostly as pyrite, while the rest was reoxidized and could potentially consume 3.0 mol O_2, i.e. one third of the calculated O_2 uptake. These redox balances are in general accordance with earlier results from Danish coastal sediments which, however, were based on oxygen uptake measurements in cores [Jørgensen, 1980].

The measured annual release of urea plus combined inorganic nitrogen from the sediment was 0.96 mol N, which could cover 29% of the nitrogen required for the measured primary production [Lomstein and Blackburn, 1992]. Sources from the land immediately surrounding Aarhus Bay and from the atmosphere contributed an additional 0.55–0.70 mol or 16–21%, the atmospheric source being one third of the terrestrial source [Sørensen and Nielsen, 1992]. The nitrate pool accumulating in the water column during winter and used by the spring phytoplankton bloom corresponded to 12–15% of the annual consumption. Over the year, an estimated 68% of the nitrogen assimilated by the phytoplankton was recycled in the water column. The total nitrogen input from sediments, land and atmosphere, ca. 1.75 mol, exceeded the loss through sedimentation, 1.06 mol, the excess presumably being exported. Dissolved organic nitrogen compounds – other than urea – were also released from the sediment at an annual rate of 0.18 mol N. The role of this DON as an N source for phytoplankton is poorly understood. The total nitrogen flux from the sediment appears to exceed the deposition minus burial rate by 25%, which may reflect accumulated inaccuracies in the many directly measured rates.

Phosphate released from the sediment was equivalent to only 16–19% of the annual plankton uptake. Land sources corresponded to only 4–9%, after a 70–80% reduction in recent years, especially due to a 90% reduction since 1990 from the largest single source, the city of Aarhus, through a new sewage treatment plant [Sørensen and Nielsen, 1992]. There has, consequently, been a notable reduction in the phosphate concentration of the water column in Aarhus Bay over the last decade, and the system now appears to be more dependent on phosphate inflow from the Baltic Sea and Kattegat. In spite of the strong reduction of the terrestrial phosphate source, phytoplankton growth still became nitrogen-limited in late summer, although the periods with potential nutrient limitation were alltogether longer. Primary production in 1990–1991 was ca. 20% lower than in 1987–89, and this decrease continued in 1992–1993. A longer time series is, however, required to discriminate this trend statistically from interannual fluctuations due to, e.g. variations in precipitation causing variations in the washout of agricultural nitrogen fertilizer. Furthermore, the problem of oxygen depletion is also coupled to conditions outside the bay, as hypoxia may originate in the heavy bottom water from Kattegat. The bottom water then becomes further oxygen depleted in the bay due to the high benthic respiration. Occasionally, e.g. in September 1990, heavy water masses from Kattegat pressed the bottom water deep into the bay and against the coast with the consequence that demersal fish were found suffocating along the shore. A bottom hypoxia can, however, also be exported from the bay when lighter Baltic water flows in and expels the bottom water to leave a uniformly oxic water column [Nielsen et al., 1993].

In conclusion, the central Aarhus Bay showed a nearly balanced carbon cycle in which more than one third of the mineralization of the primary production took place in the sediment. Nutrient regeneration and release from the sediment was a more important source for phytoplankton growth than the contributions from the atmosphere and the surrounding land. The bay system is, however, open to water and nutrient exchange with the southern Kattegat, and there has consequently been only little effect on primary productivity of the recent reductions in nutrient discharge into the bay. Especially the phosphate discharge has been strongly reduced, whereas the agricultural contribution of nitrogen has not been significantly changed. Although phosphate appeared to be the limiting nutrient for the buildup of phytoplankton biomass in late spring, inorganic nitrogen was depleted in the euphotic zone during most of the growth season.

The nitrogen and phosphorus nutrients showed important differences in the mechanisms of release from the sediment, especially in their seasonality and response to hypoxia. Nitrogen did not accumulate in the surface sediment but was gradually released throughout the year, mostly as NH_4^+ and urea. Phosphate accumulated in the top 2–3 cm of surface sediment, bound to oxidized iron minerals, and was released in pulses when hypoxia or a settling phytoplankton bloom caused a partial iron reduction in this sediment layer. It is notable, that in contrast to the open exchange of nutrients with the surrounding waters, sediment particles are largely retained within the system, and through recycling between sediment and water they are also important in regulating the nutrient balance within the bay.

References

Billen, G., C. Lancelot, and M. Meybeck. N, P, and Si retention along the aquatic continuum from land to ocean, in *Ocean Margin Processes in Global Change*, edited by R. F. C. Mantoura, J.-M. Martin, and R. Wollast, pp. 19–44, John Wiley & Sons, Chichester, 1991.

Bjørnsen, P. K., H. Kaas, H. Kaas, T. G. Nielsen, M. Olesen, and K. Richardson, Dynamics of a subsurface phytoplankton maximum in the Skagerrak, *Mar. Ecol. Prog. Ser., 95*, 279-294, 1993.

Christiansen, C., L. C. Lund-Hansen, and P. Skyum, Hydrography and material transport in Aarhus Bay (in Danish), *Havforskning fra Miljøstyrelsen, 39*, 86 pp., Danish Environmental Protection Agency, Copenhagen, 1994.

Floderus, S., Sedimentation and resuspension in Aarhus Bay (in Danish), *Havforskning fra Miljøstyrelsen, 18*, 39 pp., Danish Environmental Protection Agency, Copenhagen, 1993.

Floderus, S., and L. C. Lund-Hansen, Current related redeposition in Aarhus Bay resolved with a near bed time-series sediment trap, *Cont. Shelf Res.*, submitted.

Fossing, H., B. Thamdrup, and B. B. Jørgensen, Sulfur, iron, and manganese cycling in the sea floor of Aarhus Bay (in Danish), *Havforskning fra Miljøstyrelsen, 15*, 77 pp., Danish Environmental Protection Agency, Copenhagen, 1992.

Gundersen, J. K., and B. B. Jørgensen, Microstructure of diffusive boundary layers and the oxygen uptake of the sea floor, *Nature, 345*, 604–607, 1990.

Gundersen, J. K., R. N. Glud, and B. B. Jørgensen, Oxygen uptake of the sea floor in Aarhus Bay (in Danish), *Havforskning fra Miljøstyrelsen, 57*, 155 pp., Danish Environmental Protection Agency, Copenhagen, 1994.

Hansen, J. W., B. Thamdrup, H. Fossing, and B. B. Jørgensen, Redox balance and temperature regulation of mineralization in Aarhus Bay (in Danish), *Havforskning fra Miljøstyrelsen, 36*, 72 pp., Danish Environmental Protection Agency, Copenhagen, 1994.

Jensen, H. S., and B. Thamdrup, Iron-bound phosphorus in marine sediments as measured by bicarbonate-dithionite extraction, *Hydrobiologia, 253*, 47–59, 1993.

Jensen, H. S., P. B. Mortensen, F. Ø. Andersen, E. Rasmussen, and A. Jensen, Phosphorus cycling in a coastal marine sediment, Aarhus Bay, Denmark, *Limnol. Oceanogr. 40*, 908–917, 1995.

Jensen, M. H., E. Lomstein, and J. Sørensen, Benthic NH_4^+ and NO_3^- flux following sedimentation of a spring phytoplankton bloom in Aarhus Bight, Denmark, *Mar. Ecol. Prog. Ser.*, *61*, 87–96, 1990.

Jørgensen, B. B., Seasonal oxygen depletion in the bottom waters of a Danish fjord and its effect on the benthic community, *Oikos*, *34*, 68–76, 1980.

Jørgensen, B. B., Mineralization of organic matter in the sea bed – the role of sulphate reduction, *Nature*, *296*, 643–645, 1982.

Kemp, W. M., P. Sampou, J. Caffrey, M. Mayer, K. Henriksen, and W. R. Boynton, Ammonium recycling versus denitrification in Chesapeake Bay sediments, *Limnol. Oceanogr.*, *35*, 1545–1563, 1990.

Kruse, B. Depth distribution of the annual primary production, chlorophyll *a*, bacterial production, and O_2-respiration in a stratified Danish bay: The significance of the bottom water. *Mar. Ecol. Prog. Ser.*, submitted.

Lomstein, B. Aa., and T. H. Blackburn, Nitrogen cycling in the sea floor of Aarhus Bay (in Danish), *Havforskning fra Miljøstyrelsen*, *16*, 74 pp., Danish Environmental Protection Agency, Copenhagen, 1992.

Lomstein, B. Aa., and T. H. Blackburn, Seasonal variation in sediment carbon cycling in Aarhus Bay, Denmark, *Estuar. Coast. Shelf Sci.*, submitted.

Lund, B. Aa., and T. H. Blackburn, Urea turnover in a coastal marine sediment measured by a ^{14}C-urea short term incubation. *J. Microbiol. Meth.*, *9*, 297–308, 1989.

Lund-Hansen, L. C., M. Pejrup, J. Valeur, and A. Jensen, Gross sedimentation rates in the North Sea – Baltic Sea transition: effects of stratification, wind energy transfer, and resuspension. *Oceanol. Acta*, *16*, 205–212, 1993.

Maa, P. Y., and A. J. Metha, Mud erosion by waves: A laboratory study, *Cont. Shelf Res. 7*, 1269–1284, 1987.

Moeslund, L., B. Thamdrup, and B. B. Jørgensen, Sulfur and iron cycling in a coastal sediment: Radiotracer studies and seasonal dynamics, *Biogeochem. 27*, 129–152, 1994.

Nielsen, J., A. Lynggaard-Jensen, H. P. Hansen, and C. Simonsen, Dynamics and complexity in Aarhus Bay (in Danish), *Havforskning fra Miljøstyrelsen*, *23*, 66 pp., Danish Environmental Protection Agency, Copenhagen, 1993.

Nielsen, T. G., T. Kiørboe, and P. K. Bjørnsen, Effects of a *Chrysochromulina polylepis* subsurface bloom on the planktonic community, *Mar. Ecol. Prog. Ser.*, *62*, 21–35, 1990.

Nixon, S. W., Remineralization and nutrient cycling in coastal marine ecosystems, in *Nutrients and Estuaries*, edited by B. J. Neilson and L. E. Cronin, pp. 111–138, Humana Press, New York, 1981.

Pejrup, M., J. Valeur, and A. Jensen, Vertical sediment fluxes in the Aarhus Bay, Denmark, measured by use of sediment traps, *Cont. Shelf Res.*, submitted.

Rasmussen, H., and B. B. Jørgensen, Microelectrode studies of seasonal oxygen uptake in a coastal sediment: role of molecular diffusion, *Mar. Ecol. Prog. Ser.*, *81*, 289–303, 1992.

Richardson, K, and A. Christoffersen, Seasonal distribution and production of phytoplankton in the southern Kattegat. *Mar. Ecol. Prog. Ser.*, *78*, 217–227, 1991.

Sørensen, H. M., and K. Nielsen, Aarhus Bugt 1990–1991: *Hydrography, Nutrients and Plankton* (in Danish), 67 pp., County of Aarhus, Envir. Dept, 1992.

Suess, E., Particulate organic carbon flux in the oceans: Surface productivity and oxygen utilization, *Nature*, *288*, 260–263, 1980.

Thamdrup, B., H. Fossing, and B. B. Jørgensen, Manganese, iron, and sulfur cycling in a coastal marine sediment (Aarhus Bay, Denmark), *Geochim. Cosmochim. Acta 58*, 5115–5129, 1994.

Valeur, J. R., A. Jensen, and M. Pejrup, Turbidity, particle fluxes and mineralisation of C and N in a shallow coastal area, *Austr. J. Mar. Freshw. Res.*, *46*, 1995.

Valeur, J. R., M. Pejrup, and A. Jensen, Particulate nutrient fluxes in Vejle Fjord and Aarhus Bay (in Danish), *Havforskning fra Miljøstyrelsen*, *14*, 127 pp., Danish Environmental Protection Agency, Copenhagen, 1992.

8

Benthic Macrofauna and Demersal Fish

Lars Hagerman, Alf B. Josefson and Jørgen N. Jensen

8.1. Introduction

Eutrophication strongly affects the benthic ecosystem, because of increased primary production of planktonic microalgae or of benthic micro- and macrophytes, which are to a large extent degraded in the sediment. Since many of the macrobenthic organisms have life spans of several years, eutrophication effects will last, and sometimes be integrated, over longer time periods. The nature of the effects depends on several factors such as the amount and quality of sedimenting organic material, the degree of oxygen depletion in the bottom water, and the differential response by the benthic populations to these factors. The effects may be evident at very different levels, ranging from the physiology of individuals to the ecology of the entire benthic ecosystem. Effects of a progressing degree of eutrophication follow a sequence, starting with enhanced growth of the individuals, followed by growth in population size with associated changes in species composition, then followed by physiological and behavioral effects, and possibly with mass mortality due to anoxia as the ultimate stage [Gray 1992]. In this chapter we summarize some recent physiological and ecological studies on the effects of eutrophication on benthic macrofauna (invertebrates >1 mm) and demersal fish, with examples taken primarily from Scandinavian waters. The subjects treated are: eutrophication effects on demersal fish populations, effects of increased food levels on macrobenthos, and effects of hypoxia on macrobenthos, in particular the Norway lobster. The rationale for treating the Norway lobster in more detail is, that this economically important species has played an important role in the political debate in Denmark on eutrophication, and that it is a good representative for those infaunal benthic invertebrates which are sensitive to eutrophication.

Eutrophication in Coastal Marine Ecosystems
Coastal and Estuarine Studies, Volume 52, Pages 155–178
Copyright 1996 by the American Geophysical Union

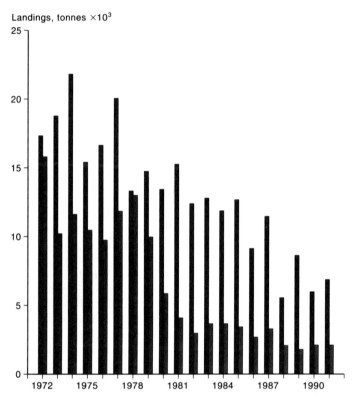

Figure 8.1. Annual landings of the demersal fish species, cod (black bars) and plaice (grey bars) from the Kattegat. Data from ICES (International Council for the Exploration of the Sea).

8.2. Effects on Fisheries

The Kattegat and Skagerrak are naturally productive marine areas which have provided the basis for an extensive fishery during at least the past 100 years. From the beginning of the 1980s, catches of the demersal fish cod and plaice, however, decreased dramatically in the Kattegat (Figure 8.1; ICES, 1992, 1993), whereas catches of dab increased in southern Kattegat [Bagge et al., 1993]. These changes were partly attributed to hypoxia, which has increased in frequency and spatial extension during the 1980s [e.g., Baden et al., 1990a; chapter 1, this vol.]. Catches of plaice and dab have been low in south-eastern Kattegat when oxygen concentrations were low in autumn [Pihl, 1989; Petersen and Pihl, 1995]. This was interpreted as the result of an avoidance reaction by the fish and/or of direct mortality. A factor contributing to the decrease in plaice and cod stocks in southern Kattegat may be the adverse effects on young stages due to eutrophication-induced mass occurrence of filamentous algae in nursery areas [Bagge et al., 1990]. Concurrent with the decrease in stock size, the individual mean size of plaice and dab has also decreased during the 1980s [Bagge et al., 1993; Petersen and Pihl, 1995]. A number of explanations related to eutrophication have been proposed to account for these changes, e.g. physiological stress, causing starvation during hypoxic periods [Petersen and Pihl, 1995] and food limitation of plaice. The abundance of food preferred by plaice, benthic bivalves and crustaceans [Blegvad, 1930], has decreased [Bagge et al., 1993], while food items used by dab, mostly brittle stars, have increased in abundance [Pearson et al., 1985; Josefson and Jensen, 1992a,b; Bagge et al. 1993]. In contrast to plaice, dab show an increasing standing stock in southern Kattegat [Bagge et al., 1993]. However, none of these hypotheses have yet been verified, and Bagge et al. [1993] suggested that food limitation is not a likely explanation for decreased growth in plaice and dab.

The important fishery for Norway lobster is also affected by eutrophication. It is known that low oxygen levels, in addition to causing starvation [Hagerman and Baden, 1988], may force the lobsters to emerge from their burrows up on the sediment surface where they are more easily caught by trawls [Bagge and Munch-Petersen, 1979]. This increased catchability is evident from data on the catch per unit effort from south-eastern Kattegat [Baden et al. 1990b], which show peak values during hypoxic conditions in autumn. The same study indicates a local drastic decrease in the lobster stock, with zero catch in 1989, whereas fishery statistics do not indicate similar changes over a larger area in the eastern Kattegat [Baden et al., 1990b]. Fishing for lobster in southern Kattegat became unprofitable and nearly stopped at the end of the decade [Baden et al., 1990a] and the fishery moved northward, which in itself may indicate a general decrease of the stock.

The most negative effects of progressing eutrophication on fisheries of demersal fish are apparent in the Kattegat, whereas few effects have been reported from the Skagerrak, and they are at least not apparent from the data on fish landings during the period 1972–1987 when combined for Skagerrak and Kattegat (ICES data in Josefson [1990].

To conclude, although direct field evidence for effects of eutrophication on the fishery exists from local areas and although several likely mechanisms have been proposed for this, the fishery statistics suggest different trends in the Kattegat and in the Skagerrak which at present are difficult to explain only with overfishing.

8.3. Effects of Increased Food Levels on Macrobenthos

As a consequence of eutrophication, an enhanced sedimentation of phytoplankton to the sea floor will increase the food supply to the benthic fauna. Unless factors such as hypoxia and available physical space are limiting, the benthic fauna is likely to respond to an increased food level by increased abundance and biomass (standing stock). Among the underlying assumptions for such a response are, that the macrobenthos is normally food-limited, and that mortality rates remain unchanged. Examples given in the following from the Skagerrak-Kattegat are mostly from benthos of offshore soft sediments, where physical space is unlikely to be a limiting factor [Peterson, 1979].

Several studies in adjacent sea areas have demonstrated increased abundance and/or biomass in recent years, some of them attributing the changes to eutrophication: the Baltic Sea [Cederwall and Elmgren, 1980; Brey, 1986], the Dutch Wadden Sea [Beukema and Cadee, 1986], and the central North Sea [Kröncke, 1992].

Although it is well known that macrobenthos may respond to increased organic load by increasing the standing stock [Pearson and Rosenberg, 1978], little is known about the mechanisms involved. Questions to be further addressed are: Which components of the organic material are most important as food, how is phytoplankton transported to the benthos, what are the most important response mechanisms of the benthos, what is the expected response time, which faunal components are favored/disfavored?

Benthic species may respond to increased food levels by increased individual growth rate and/or increased number of individuals [Dauer and Conner, 1980]. The latter response may involve increased survival of juveniles or increased reproductive fitness [Levin, 1986], thus leading to an increased number of recruits. While growth of the resident animals will start soon after a phytoplankton sedimentation pulse has reached the bottom [Graf et al., 1982; Christensen and Kanneworff, 1985], effects on recruitment will be evident only some time after the food input, typically after a year or more [Johnson and Wiederholm, 1992]. Which of the two mechanisms of biomass increase will dominate will depend on factors such as the time of food input to the bottom relative to the growth potential and stage in the reproductive cycle of the benthic populations, the magnitude of the input, etc.

8.4. Effects on Benthos in the Kattegat-Skagerrak Area.

The great economic importance of fishery after demersal fish and the fact that this fish category utilizes macrobenthic species as food [Blegvad, 1930] were the background for an extensive mapping of benthos in Danish waters in the beginning of this century [Petersen, 1913, 1915]. The result was a unique material, a database, from a relatively pristine time period, against which the recent status of the benthos can be compared. Comparisons of the faunal biomass and species composition in the 1980s with the situation in the beginning of the century in the Kattegat-Skagerrak [Pearson et al., 1985; Rosenberg et al., 1987; Josefson and Jensen, 1992a] have shown marked differences (Figure 8.2). Lower biomasses of the total macrofauna in the 1980s were observed at shallow depths of <50 m, while increases were dominant at greater depths. At intermediate depths both decreases and increases were observed. Sea urchins (*Echinocardium cordatum*) and some molluscs had decreased whereas brittle stars and polychaetes had increased their biomass. The most marked increase, >10-fold, of an individual species was found for the filter feeding brittle star, *Amphiura filiformis*, in the southern Kattegat. Generally, there was a shift from large to small species and a decrease in individual size was also apparent within the dominating species, *Amphiura filiformis* [Josefson and Jensen, 1992a]. Rosenberg and Möller [1979] compared the fauna in Swedish fjords during 1930–

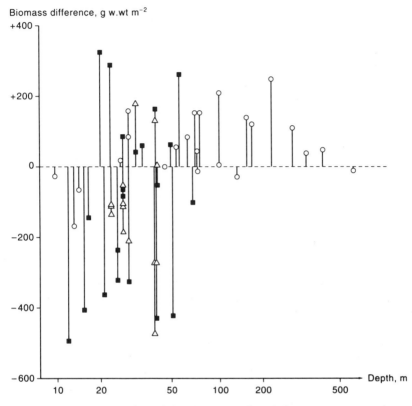

Figure 8.2. Comparison of total benthic macrofaunal biomass (raw weight) in the Kattegat–Skagerrak between the 1980s and 1910s plotted against water depth on a log scale. Data from Pearson et al. [1985], squares; Rosenberg et al. [1987], circles; Josefson and Jensen [1992a], triangles. + indicates increase and – indicates decrease since the 1910s.

1970, before hypoxia-induced problems became apparent in the Kattegat, and showed increased biomass of the total macrobenthos in the 1970s irrespective of depth.

Time series of benthos measurements with at least annual sampling from the period 1973–88 in the eastern Skagerrak and northern Kattegat [Josefson, 1990] suggest a doubling of the total macrobenthic biomass during this period, with the fastest increase in the late 1970s and early 1980s [Austen et al., 1991; Josefson et al., 1993]. In most places the biomass increase was accompanied by an increase in the total density of individuals [Josefson et al., 1993]. With a ca. one-year time lag at 100 m, and two years at 300 m, biomass, and in most instances also density, measured in spring, were significantly correlated with runoff from surrounding land areas, which determines the inorganic N-nutrient pool for the spring phytoplankton bloom and the annual mean chlorophyll-*a* concentrations in the euphotic zone [Josefson et al., 1993; chapter 9, this vol.]. Examples of the relation between water

Biomass, g w.wt 0.1 m^{-2}

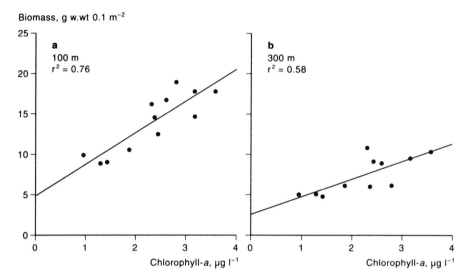

Figure 8.3. Benthic biomass in spring at 100 m (a) and 300 m (b) depth plotted against mean annual chlorophyll-*a* in the euphotic water column, one (a) and two (b) years earlier, during the period 1973–1988. Based on data in Josefson et. al [1993].

column chlorophyll and biomass from 100 and 300 m water depth are shown in Figure 8.3. The changes of the benthos during 1973–1988 involved changes in species composition and in dominance patterns [Austen et al., 1991; Josefson et al., 1993]. The greatest increases in dominance were shown by ophiuroids, notably *Amphiura filiformis*, and polychaetes [Josefson, 1990], species which were already present among dominants at the beginning of the period.

A study of somatic growth in the one species contributing most to the increase in total macrobenthic biomass, *Amphiura filiformis*, showed that growth rates were at least equally high in the beginning of the 1980s at high densities and large individual size compared to the 1970s, suggesting high energy input to the bottom during this period [Josefson and Jensen, 1992b].

The general doubling of biomass in Kattegat was of the same magnitude as for chlorophyll-*a* concentrations in the euphotic zone [Josefson et al., 1993] and for primary production (e.g., Richardson and Heilmann, 1995). Studies of sedimentation in the Kattegat suggest sedimentation times of one to several weeks for the phytoplankton biomass to reach the depths of the present area (chapter 4). A two-year time lag between the changes in the pelagic and the benthic response (when sampling benthos in spring) is reasonable, provided that the dominating mechanism of increase is an increased recruitment due to increased reproductive fitness of residents. A similar lag between changes in benthic biomass or abundance and pelagic chlorophyll levels has been reported from the western North Sea [Buchanan, 1993] and is evident in time series from the southern North Sea [Beukema, 1991; Beukema and Cadee, 1991]. The described biomass increase has occurred in an area which, in contrast to southern Kattegat, seldom experiences hypoxic conditions in the bottom water. The biomass increase may thus be explained by increased food input to the benthos in the 1980s. The alternative explanation involving decreased mortality is less likely because the greatest macrobenthic biomass increase occurred in areas from which landings of potential predators, the demersal fish, have remained at the same level. It should be pointed out, however, that other explanations in addition to eutrophication have been proposed for benthic changes in the Skagerrak and the North Sea. For instance, Evans and Edwards [1993] presented evidence of climate-related effects in the pelagic system in the western North Sea, which in turn could have resulted in increased input of organic matter to the sea floor in the 1980s.

8.5. Adaptational and Physiological Effects of Hypoxia

Benthic invertebrates exposed to low oxygen tensions show different responses due to their different behavioral strategies and physiological tolerances. Epibenthic species have mostly a low tolerance to decreased oxygen tensions. Crustaceans such as *Crangon crangon*, *Palaemon adspersus*, and *Carcinus maenas* are examples of this. Most species show initial reactions to hypoxia when the oxygen tension decreases below half air saturation and, just like most fish, try to escape from the hypoxic area by walking away or swimming horizontally or vertically. They may also move up from the sediment on to stones or algae where the oxygen availability is often higher. Animals that cannot escape react to hypoxia with other behavior or, if the severeness of hypoxia increases, with physiological and biochemical mechanisms. The most stationary species, which frequently have to cope with hypoxic conditions in their habitat, show mostly better tolerances and have developed more efficient mechanisms to conserve energy as long as possible.

It should be pointed out, that it is the partial pressure and the availability of oxygen, rather than the concentration of oxygen in the immediate environment, that determines whether an organism can survive hypoxia. Thus, flowing water can increase the oxygen availability considerably for a stationary benthic animal relative to stagnant water. This is especially true for smaller animals lacking ventilatory organs. As long as the supply of oxygen can match the consumption, the animal will not suffer from hypoxia, even though the oxygen content in the water may be very low. Experiments with smaller animals, such as juvenile *Amphiura*, in which water is pumped through a respiratory chamber, will thus tend to overestimate the in situ tolerance to hypoxia since the experimental animal is continuously provided with more oxygen than under natural conditions. The critical oxygen content under such experimental conditions may even be so low that it is difficult to determine.

The most common hypoxia response used by species with ventilatory organs is to somehow increase the oxygen availability. Burrowing polychaetes and most crustaceans increase their ventilatory rate when oxygen tensions fall below half air saturation. They can in this way irrigate and oxygenate their burrows and maintain their level of aerobic respiration. They behave as oxyregulators [Mangum, 1973].

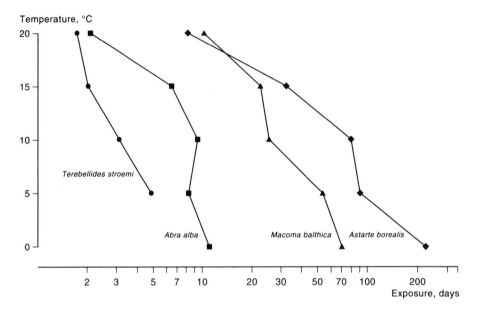

Figure 8.4. Effect of temperature on the resistance (LT_{50}) to anoxic conditions in some benthic invertebrates: a polychaete, *T. stroemi*, and three bivalves. After Dries and Theede [1974].

The tolerance to completely anoxic conditions shows wide variation among different species. An active crustacean, such as *Crangon crangon*, has an LT_{50} (= the time until 50% of the animals have died) in anoxia at 10°C of only two hours. At the other extreme, the bivalve, *Astarte borealis*, from the western Baltic has an LT_{50} of 2000 hours [Dries and Theede, 1974]. Bivalves generally have the highest tolerances, followed by polychaetes, e.g. *Nereis diversicolor*: LT_{50} 100 hours, and *Terebellides stroemi*: LT_{50} 90 hours. Crustaceans have the shortest survival times. One exception is the Baltic isopod, *Saduria entomon*, which can survive up to 200 hours of anoxia [Hagerman and Szaniawska, 1990]. The presence of H_2S reduces the survival of several species under hypoxic or anoxic conditions. On the other hand, a high tolerance to anoxia is often combined with a tolerance to H_2S [Theede et al., 1969; Vismann, 1991]. This is important in the habitat, as anoxia is mostly, if not always, followed by an increased upward diffusion of H_2S from reduced sediment layers (chapter 6). The high tolerance is mainly due to mechanisms at the cellular level. A sufficient content of metabolic storage substrate (e.g. glycogen) and the ability to reduce the overall metabolic requirements by reducing different types of activity are necessary to achieve this high tolerance. Low mobility is often correlated with a low metabolic rate and thus a high tolerance, e.g. in sediment-dwelling bivalves, such as *Arctica islandica*. Among related species, those with a higher metabolic rate and smaller energetic reserves will generally have a shorter survival time. Anoxia and H_2S are tolerated better at low temperatures, and the tolerance decreases steeply with increasing temperature (Figure 8.4) [Theede, 1984; Hagerman and Szaniawska, 1992]. This suggests that species with a high metabolic rate and small energy stores will have a low anoxic tolerance and thus a short survival time.

Echinoderms have always been considered to be sensitive to low oxygen tensions [Gaston, 1985], probably due to their relatively large size combined with a lack of real ventilatory appendages. As already mentioned, they are quantitatively important, especially the brittle stars, on or in soft bottoms of the Kattegat region. Rosenberg et al. [1991] observed that at around 10% of air saturation, *Amphiura* emerged from the sediment, a behavior that increases the oxygen availability considerably for the animal. It is, however, difficult under the experimental conditions used, to conclude about the tolerance of *Amphiura* to low oxygen tensions in situ. Other echinoderms, such as *Echinocardium cordatum*, which are normally burrowed deep within the sediment, seem to be the first species to escape from the sediment upon anoxia, and mass mortality has been reported even at 30% air saturation of oxygen [Detlefsen and Westerhagen, 1984]. *E. cordatum* harbors sulfide-oxidizing bacteria in the intestine [Temara et al., 1993], which indicates the presence of sulfide in the intestine. *E. cordatum* thus probably requires a sufficient oxygen tension for a bacterial detoxification of sulfide and flees from anoxic sediments to avoid excessive sulfide concentrations.

The ability to withstand extensive periods of low oxygen concentration has been demonstrated in several laboratory experiments. It may therefore appear astonishing that apparent effects of oxygen depletion can be detected in the field at comparatively higher levels of oxygen concentration. This discrepancy might be explained as an artifact in the monitoring of oxygen. Measurements of bottom water concentrations of oxygen are mostly done at a height of 0.5–1 m above the bottom, and may not be representative for the oxygen concentration just above the sediment. Behavioral responses of benthic animals experiencing hypoxia indicate that a sharp oxygen gradient may exist just above the sediment. Thus, Jørgensen [1980] observed that bivalves extended their siphons up into the water during hypoxia.

Also predator-prey relationships are affected by oxygen depletion. Laboratory experiments [Sandberg, 1994] indicate that the predatory isopod, *Saduria entomon*, predated less efficient on a mobile species of amphipod during hypoxia than on another less mobile amphipod species. Changes in predation rate may alter the population size and dynamics of the prey species and may especially in species-poor communities cause great changes at the community level.

Facultative anaerobiosis is a general phenomenon among benthic soft-bottom species but the metabolic pathways used in anaerobiosis can vary. Anaerobiosis occurs only when the oxygen tension has decreased to a level where basic metabolic needs can no longer be maintained aerobically. It can be either an exercise anaerobiosis or an environmental anaerobiosis [de Zwaan and Putzer, 1985]. The classical glycolysis with lactate (lactic acid) as an anaerobic endproduct is found in active species, such as crustaceans and some errant polychaetes. Other groups, mainly polychaetes and bivalves, accumulate fatty acids such as succinate and, if the period of anaerobic metabolism is prolonged, sometimes also propionate. The increase in

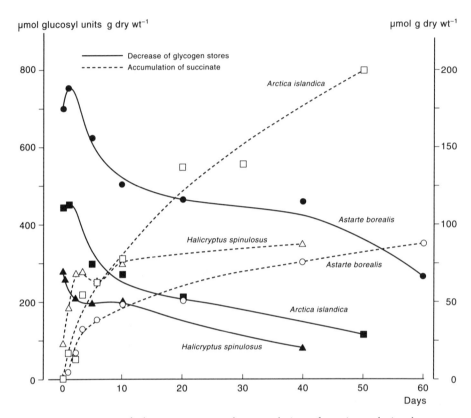

Figure 8.5. Decrease of glycogen stores and accumulation of succinate during long-term anaerobiosis in some benthic invertebrates: a priapulid, *H. spinulosus*, and two bivalves. After Oeschger [1990].

succinate is accompanied by a decrease in glycogen, for example in the bivalves, *Arctica islandica*, *Astarte borealis*, and the priapulid, *Halicryptus spinulosus*, which are common in soft bottoms in the western Baltic and Kattegat (Figure 8.5) [Oeschger, 1990]. During an initial phase of short-term anaerobiosis, aspartate is used as a substrate giving opines and acetate or alanine as endproducts. Some of the more active polychaetes, such as *Nereis diversicolor*, also produce lactate during short-term anaerobiosis, for instance during low tide [Schöttler et al., 1984]. Low salinity will under severe hypoxia or anoxia result in an increased stress and a reduction in survival time. An osmotic adjustment to lower salinity requires energy and the drain on energy stores will be faster, thus resulting in a faster accumulation of anaerobic endproducts as shown for some polychaetes (e.g. *Arenicola marina*) by Schöttler et al. [1990] (Figure 8.6). The burrowing polychaete, *Nephtys hombergii*, is rather tolerant to anoxia. It survives at least five days at 12°C, and the animal also remains active with an aerobic metabolism (but possibly with functional anaerobiosis) down to 15–20 torr O_2 and partly aerobic even at 7 torr [Schöttler, 1982]. No lactate is, however, produced in *N. hombergii*, but instead opines seem

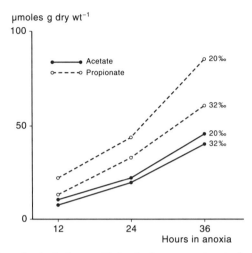

Figure 8.6. Excretion of volatile fatty acids by the lugworm, *Arenicola marina*, after different times of anaerobic exposure and at different salinities. After Schöttler et al. [1990].

to be the anaerobic endproducts at the beginning and succinate later during anaerobiosis (Schöttler, op. cit.). The high capacity of active, burrowing animals for maintaining an aerobic metabolism during severe hypoxia is probably a characteristic phenomenon. Tube-dwelling animals have a lower capacity, as they can compensate for low oxygen by increasing the ventilation rate and thus bring water from above directly into their tubes.

8.6. The Norway Lobster – a Case Animal

The Norway lobster, *Nephrops norvegicus*, is economically very important in the Kattegat-Skagerrak area. The progressing hypoxia during the last decades has had a detrimental impact on the stock of this species and has motivated a Danish-Swedish-British collaborative research on the ecophysiology of *N. norvegicus*. Norway lobsters, which were dead or dying due to hypoxia, were found for the first time in autumn 1985. Oxygen deficiency was serious in 1988 (< 15% air saturation), lasted longer, and reached an extension of 4000 km². No *Nephrops* were caught after that period. *N. norvegicus* lives on and in mud-clay bottoms of a consistency which makes it possible for the animals to excavate horizontal burrows. They use their first three pairs of pereiopods and third maxillipeds to excavate and transport heaps of sediment to the openings [Atkinson and Taylor, 1988]. The burrows can be of several types ranging from a single opening with only one inhabitant in the shaft to burrows constituting a network with many openings and more than one inhabitant. Most common is a single burrow ("U-tube") with two openings [Rice and Chapman, 1971]. Typically, one opening is broad and craterlike (the entrance) with a heap of excavated material in front, while the other opening is simple [Dybern and Høisæter, 1965]. The animals use the front opening for both exit and entrance [Figureido and Thomas, 1967]. The burrows often occur in small groups. The most important function of the burrows is most certainly protection from predators. Other functions might be as moulting site and as the center of territory. *N. norvegicus* leaves the burrow for foraging and for mating. Some food might be brought back to the burrow for feeding. The burrow thus serves as the "home" for an individual. Juveniles often inhabit small tunnels connected to the burrows of the adults [Chapman, 1980]. It is mostly males that leave their burrows for excursions and under normoxic conditions females stay in the burrows during the egg-carrying period (May-September) [Figureido and Thomas, 1967]. It is only when hypoxic autumn conditions force the females to emerge that they are caught in greater numbers. *Nephrops* is actively ventilating its burrow. By intermittent beating of the pleopods, a current of water is circulated through the burrow, thus preventing it from becoming too hypoxic (provided the water outside is more or less normoxic). Observations by Atkinson and Taylor [1988] indicate that gill ventilation of the scaphognathites can be sufficient to prevent hypoxic burrows under normoxic conditions in the overlying water. It is also a possibility that a bottom current can help a passive exchange of water in the burrow (cf. Huettel and Gust [1992]).

Experimental investigations have shown that when the oxygen concentration decreases to ca. 50% of air saturation, *Nephrops* rise higher on their legs, show signs of restlessness, and markedly increase their cleaning behavior. At even lower oxygen tensions, they emerge from their burrows [Hagerman and Uglow, 1985] and at 10–20% of air saturation feeding activity stops and they remain inactive on stretched legs with lowered claws, if possible on top of a small mound, thus trying to reach more oxygen-rich surroundings [Hallbäck and Ulmestrand, 1990]. It is clear that *Nephrops* will be more vulnerable to trawl catch under these conditions, obviously an explanation for the increased autumn catches in the early 1980s, which occurred at the same time as severe hypoxia.

A useful reaction to hypoxia is, as mentioned above, to try to increase the availability of oxygen by increasing the water flow to the gills. The scaphognathite beat frequency of quiescent *Nephrops* increases from a normoxic rate of ca. 60 min^{-1} to ca. 120 min^{-1} at an oxygen tension of 40 torr. The heart rate remains stable over the same range of oxygen tensions (Figure 8.7) [Hagerman and Uglow, 1985]. This must be regarded as an acute response to short-term hypoxia. Scaphognathite rates will decrease below normoxic rates if hypoxia prevails for weeks [Kwee, 1993]. In common with other decapods, *Nephrops* shows no ventilatory pauses or cardiac arrests, which are common during normoxia, when subjected to moderate short-term hypoxia.

Increased ventilation increases the extraction efficiency for oxygen (from 20–30% to 30–40%) and this is important for the ability to maintain stable O_2 uptake levels during hypoxic events. Hagerman and Uglow [1985] calculated that, at half air saturation, 0.25 ml water was pumped per scaphognathite beat per gill, or 75 μl O_2 was pumped past the gills each minute. The extraction efficiency compares well with an oxygen uptake rate of 15 μl g^{-1} h^{-1} for a *N. norvegicus* of 100 g wet weight.

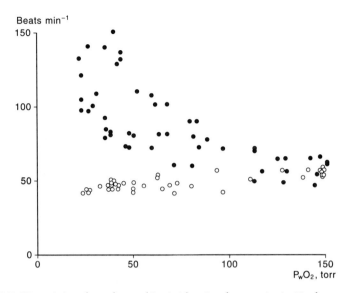

Figure 8.7. Heart (○) and scaphognathite (●) beating frequencies in *Nephrops norvegicus* as a function of water oxygen tensions (10°C, 30–32 psu). After Hagerman and Uglow [1985].

The most sensitive periods of decapod life, from an ecological as well as from a physiological viewpoint, are the moult and immediate postmoult stages, when the new carapace is still soft. The individual is during this stage an easy prey, so moulting must take place in a protected place, i.e. in the burrow. Cuticle permeability is increased and so is the metabolic demand, which is shown as a considerable increase in oxygen consumption rate during the moult and immediate postmoult [Alcaraz and Sardà, 1981]. It is thus obvious that exposure to hypoxia during moulting can be disastrous to the individual. Juvenile lobsters moult several times a year, while adults moult once a year [Figureido and Thomas, 1967].

Most decapods, and especially the burrowing species, keep their respiratory rate stable over a wide range of oxygen tensions. The tension where stable oxygen uptake is no longer possible is called the P_c or critical oxygen tension. For *N. norvegicus* this tension has been found to be ca. 40 torr [Hagerman and Uglow, 1985]. Down to this tension it is possible for the species to maintain an active life, feeding etc., whereas below the P_c the activity is decreased, feeding stops, and an anaerobic metabolism may start.

Survival at oxygen tensions below the P_c is dependent on duration and, the more severe the hypoxia, on energy production via an anaerobic metabolism. In *N. norvegicus*, like in most other decapods, glycolysis is the major anaerobic metabolic pathway [Gäde, 1983]. Glycogen is used as a substrate and via glucose- and fructosephosphate, phosphoenolpyruvate and pyruvate, lactate (lactic acid) is produced intracellularly and leaked to the blood where it accumulates and can be measured. If the animal is suddenly exposed to hypoxia, lactate accumulates at oxygen tensions below 15 torr (10% of air saturation). If the animal is gradually adapted to hypoxia over a 30-day period, accumulation starts only at 9–12 torr [Schlütter, 1993]. The animals survive for only two days at 15 torr [Hagerman et al., 1990]. Obviously, an anaerobic metabolism starts at oxygen tensions far below the P_c and, by being totally inactive, N. norvegicus can reduce its metabolic rate and utilize the available oxygen in the interval 40–15 torr, which is energetically more efficient than turning to an anaerobic metabolism. The accumulation of lactate is reversible and upon return of oxygenated conditions lactate can be metabolized in the hepatopancreas, which is accompanied by an increased oxygen uptake (oxygen debt) [Bridges and Brand, 1980]. Even if the animal is slowly adapted to severe hypoxia, the survival time is very limited and once severe hypoxia (< 10% of air saturation) has developed it is a question of only a few days before the population of N. norvegicus dies.

A consequence of the initial hyperventilation during moderate hypoxia is an increase in blood (haemolymph) pH, i.e. a respiratory alkalosis. The increase is from a normoxic pH of 7.75–7.90 to a hypoxic pH of 7.90–8.10 at P_{O_2} 60 torr [Hagerman and Uglow, 1985]. The increased ventilation changes the blood acid-base balance by a wash-out of CO_2, which decreases the haemolymph P_{CO_2} (Truchot, 1983). Upon prolonged exposure to hypoxia and especially to severe hypoxia a respiratory acidosis occurs. This is mainly due to the accumulation of lactate.

Like most other crustaceans, Nephrops norvegicus has the copper-containing haemocyanin as respiratory blood pigment. The haemocyanin is not bound in erythrocytes but is dissolved in the blood. Haemocyanin shows a considerable variation among individuals depending on their feeding level and moult stage. Starvation is known to decrease decapod haemocyanin concentration [Uglow, 1969; Hagerman, 1986] and the quality of food influences the final level [Hagerman, 1983]. It has experimentally been shown that moderate hypoxia (half-saturation) induces a synthesis of haemocyanin in N. norvegicus. Despite a large individual variation, biosynthesis averaged 8% of the original value per day, i.e. it took about two weeks to double or replace the whole haemocyanin pool. In N. norvegicus exposed to 60 torr (approx. half-saturation) for 12 days, haemocyanin concentration increased from the normoxic value of 0.28 mmol l^{-1} to hypoxic 0.59 mmol l^{-1}, whereas controls showed no changes [Hagerman and Uglow, 1985]. This synthesis is supposed to serve a compensatory function for the lowered oxygen tension of the ambient water, but the response requires an appropriate feeding level. Starved crustaceans show no such synthesis of haemocyanin but can instead use the haemocyanin as a protein resource with a very low haemocyanin level as the result [Hagerman, 1986]. At oxygen tensions below 30 torr, the haemocyanin concentration decreases rapidly. N. norvegicus uses its haemocyanin for metabolic needs when its stops feeding (hypoxia-induced starvation) [Schlütter, 1993].

Very low haemocyanin levels (0.03 mmol l^{-1} blood as compared to 0.5–0.8 mmol l^{-1} for well-fed normoxic individuals) were found in N. norvegicus collected in areas subjected to severe hypoxia (13–22 torr) in southern Kattegat in the autumn 1986 [Hagerman and Baden, 1988]. Such low haemocyanin levels make N. norvegicus even more vulnerable to hypoxia. Haemocyanin is especially under hypoxic conditions the important carrier of oxygen to the tissues and a lower concentration of haemocyanin transports less oxygen, even if it functions in a normal way.

Autumn storms normally reoxygenate bottom waters. This was also the case in southern Kattegat in 1986 and one month after reoxygenation the mean haemocyanin concentration in *N. norvegicus* had increased to 0.76 mmol l^{-1}, a daily synthesis rate of 0.024 mmol l^{-1} blood as also shown in the earlier experimental studies [Hagerman and Uglow, 1985; Hagerman and Baden, 1988]. The recovery was, however, very slow or even absent in some areas. Feeding experiments have shown that only decapods which have other crustaceans as important food items can synthesize haemocyanin to optimal levels [Hagerman, 1983; Depledge and Bjerregaard, 1989; Baden et al., 1990b]. The content of crustaceans in the stomachs of *N. norvegicus* from southern Kattegat after reoxygenation of the bottom water was only $\frac{1}{5}$ lower than in those from northern Kattegat. Severe hypoxia means starvation not only to *N. norvegicus* but also to other benthic crustaceans as well as other invertebrate groups. As food is the most important source of essential proteins, vitamins and minerals (incl. copper), it is possible that a mineral deficiency might have prevented the posthypoxic biosynthesis of haemocyanin in some areas [Baden et al., 1990b]. It has also been noted that manganese is leaking from the sediment during severe hypoxia [Garringa et al., 1991; Thamdrup et al., 1994]. This manganese can be deposited as manganese dioxide on the gills and carapace of *N. norvegicus*, and it was observed that, as copper concentrations fell in some tissues, manganese concentrations markedly increased [Baden et al., 1994]. It is thus a possibility needing further study that manganese might interact with copper during recovery from hypoxia and influence biosynthesis of haemocyanin.

The concentration of haemocyanin indicates indirectly the amount of oxygen that can be bound to the pigment, but says nothing about at what oxygen tension the binding can and will take place. This oxygen affinity is expressed by the p_{50}, the oxygen tension where 50% of the haemocyanin is saturated with oxygen. The affinity for oxygen can vary with the environmental conditions [Mangum, 1983] and often so that species subjected to hypoxia can take up oxygen more easily from the water, i.e. they have a higher affinity for oxygen. Burrowing species also tend to have a higher affinity for oxygen [Bridges, 1986]. *N. norvegicus* adapted to normoxic conditions were found to have a P_{50} of 11.5 torr at pH 7.9 [Bridges, 1986] and a slightly, but not significantly, higher affinity in hypoxia-adapted individuals [Schlütter, 1993]. The oxygen affinity can be modified by several extrinsic and intrinsic factors. One such intrinsic factor is the lactate concentration in the haemolymph. Lactate can improve the oxygen affinity and this improvement can counteract the decrease in affinity caused by the lactate-induced lowering of pH [Truchot, 1980]. In *N. norvegicus*, haemolymph lactate improved P_{50} ca. 3 torr, i.e. from 11.5 to 8.5, and this is similar to what has been found for other decapods [Zeis et al., 1992]. Also urate (uric acid), which accumulate during hypoxia lower than ca. 25 torr, can modify the oxygen affinity of haemocyanin [Lallier et al., 1987]. The ecological consequence of the modified affinity is a better oxygen uptake at severe hypoxia, which will enable *N. norvegicus* to utilize an aerobic metabolism longer before an anaerobic metabolism is totally taking over, i.e. an increasing hypoxic tolerance.

Anoxic sediments rich in decomposing organic material often contain toxic hydrogen sulfide. When reaching the oxic zone, hydrogen sulfide is oxidized to thiosulfate and sulfate (chapter 6). When severe hypoxia occurs, hydrogen sulfide can diffuse out into the burrow water or the near-bottom water. Some macrobenthic species, generally those that may be temporarily exposed to hydrogen sulfide in the environment, have developed tolerances and detoxification mechanisms [Vismann, 1990; Vismann, 1991]. It is possible that N. norvegicus in certain areas is exposed to hydrogen sulfide from the sediment during severe autumn hypoxia, but it is not known whether they have developed such mechanisms.

The physiological ecology and adaptations to hypoxia of the Norway lobster are typical for the larger benthic crustaceans. A species showing extremely good adaptation to hypoxia and even anoxia is the Baltic digging isopod, *Saduria entomon*, which has an anaerobic metabolism with alanine or lactate as endproducts, depending on inactivity or activity, respectively, during the anoxic exposure. It has also the ability to extract oxygen very efficiently from the water due to a very high oxygen affinity of the haemocyanin. *S. entomon* shows an excellent ability to detoxify hydrogen sulfide (Vismann 1991, Hagerman and Vismann 1993). The ecological physiology of *S. entomon* was reviewed by Hagerman and Szaniawska (1992).

8.7. Effect of Hypoxia on Macrobenthic Communities

Severe seasonal hypoxia and anoxia may lead to areas of the sea floor totally devoid of macrofauna, yet this situation is rare. Such "dead bottoms" without macroscopic life have been observed occasionally in North European waters [Josefson and Widbom, 1988; Fallesen, 1992] and, in particular, in parts of the Baltic Sea [Weigelt and Rumohr, 1986; Andersin et al., 1990].

Reductions in species diversity, abundance and biomass due to hypoxia and anoxia have been observed in parts of the North Sea, the Kattegat, and the Baltic Sea [Weigelt and Rumohr, 1986; Westernhagen et al., 1986; Rosenberg and Loo, 1988; Andersin et al., 1990; Niermann et al., 1990; Josefson and Jensen 1992a; Rosenberg et al., 1992]. Parts of the German Bight were affected by oxygen depletion in the period 1981–1983 and very low oxygen concentrations in 1983 (< 1 mg O_2 l^{-1}) caused a reduction in species diversity, abundance and biomass [Niermann et al., 1990]. Recovery took place during the following years and in 1986 the "prehypoxic" benthic fauna had been reestablished.

A severe and wide ranging oxygen depletion in Kiel Bay, western Baltic Sea, in 1981 resulted in a decrease in biomass, abundance and species diversity below the halocline [Weigelt and Rumohr, 1986]. Only a few species survived, notably the bivalves *Arctica islandica*, *Astarte* spp., and *Corbula gibba*, and the sipunculid, *Halicryptus spinulosus*. These are among the species which in laboratory experiments have been particularly resistant to hypoxia and anoxia. A review of the macrofauna data from Kiel Bay indicates a shift from larger to smaller mean size of individuals [Weigelt 1991].

Hypoxia and anoxia may result in a shift in species distributions. Fallesen and Jørgensen [1991] observed a shift in the distribution of the polychaete species, *Nephtys hombergii* and *Nephtys ciliata*, following a severe oxygen depletion in Århus Bay, Denmark, in 1981. Before the oxygen depletion, the distributions of the two species were complementary due to differences in their sediment preference, while after the oxygen depletion, *N. hombergii* dominated the whole area as a result of their successful recolonization in 1982.

Low concentrations of oxygen (< 1 ml l^{-1}) were observed in parts of the southern Kattegat in autumn 1988 [Ærtebjerg et al., 1990; Rosenberg et al., 1992]. Mass mortality of *Abra alba* and reports of large numbers of dead invertebrates captured in demersal fish trawls in south-eastern Kattegat were coupled to this event [Baden et al., 1990]. Stations sampled in 1911/1912 by Petersen [1913] and in 1984 by Pearson et al. [1985] were revisited in order to detect potential effects of the 1988 hypoxia event on the benthic macrofauna [Josefson and Jensen, 1992a]. The results indicated reductions of biomass in 1989 compared to 1984, whereas the abundance was unchanged. The greatest biomass reductions of 70–80% were found in areas where the oxygen concentration was below 1 ml l^{-1} in autumn 1988 (Figure 1.6). The decreases in biomass were most pronounced at depths just below the halocline at 15 m water depth. An explanation for this may be that the initial pool of oxygen available for the oxidation of phytoplankton sinking below the halocline, which acts as a barrier to the exchange of oxygen, is proportional to the height of the water column below the halocline. Rosenberg and Loo [1988] and Rosenberg et al. [1992] also fund the largest decrease in biomass at stations just below the halocline following the hypoxic event in autumn 1988 in south-eastern Kattegat, where sensitive species were eliminated, thus leaving hypoxia-resistant species such as the bivalve, ocean quahog (*Arctica islandica*).

The results of a 1989 revisit [Josefson and Jensen, 1992a] confirmed the general differences at the species level found in 1984 as compared to the situation in 1911–1912. The biomass and abundance of the brittle star, *Amphiura filiformis*, were still much higher and the biomass of the echinoid, *Echinocardium cordatum*, was still lower than in the beginning of the century. A previously observed decrease in the mean individual weight of the total fauna [Pearson et al., 1985] has continued, and the mean weight was on the average 50% lower in 1989 compared with 1984. Several phyla of invertebrates contributed to this decrease, although especially ophiuroids (primarily *Amphiura filiformis*) and polychaetes decreased in size. The change toward smaller-sized animals may result from hypoxia-induced elimination of older individuals and ensuing recruitment of juveniles.

A comparison of benthos data from the beginning of this century with present monitoring data in southern Lillebælt, Denmark, suggests that the areal extension of azoic bottoms has increased and that species which are sensitive to low oxygen concentrations have decreased throughout the area [Leonhard & Varming, 1992] (Figure 1.6).

8.8. Conclusions

Both the synoptic comparisons over 80 years and time series from the last two decades suggest an increase in benthic macrofaunal biomass at water depths greater than ca. 50 m in the northern part of the Kattegat and eastern Skagerrak. At depths shallower than 50 m, mainly in the Kattegat, reductions of biomass have occurred in several places. This is also the area from which adverse effects on fisheries have been reported.

How do the observed changes of macrobenthos populations fit with predictions from previous models of eutrophication-induced responses in the benthos? Existing models by Pearson and Rosenberg [1978] and Rhoads and Boyer [1982] allow qualitative predictions of successional patterns in relation to changed organic load. For instance, they predict dominance of small-sized species or individuals and dominance of species feeding at the sediment-water interface in contrast to deep-burrowing forms, in areas affected by eutrophication. Indeed, such changes seem to have occurred in the Kattegat. It is not known, however, to what extent intensive trawling for fish and lobster have contributed to this change, since trawling may be expected to affect size in a similar manner as eutrophication [e.g., Graham, 1955].

The Pearson and Rosenberg [1978] model describes an increased biomass early in the eutrophication process, whereas later, when oxygen conditions deteriorate, the biomass will decrease and eventually the original species will be replaced by opportunists. When looking at the Kattegat-Skagerrak area as a whole, the changes in the benthos appear to conform to this model. Generally, the energy input to the benthos is inversely related to the water depth and directly related to the phytoplankton productivity in the overlying water [Suess, 1980]. Thus, when the level of nutrient enrichment increases, which has occurred during the last decades, we would expect a response in the benthic community which depends on the water depth and the distance to the main nutrient sources. In deep areas, distant from the productive zone, the biomass generally showed an increase, whereas in shallow areas, some of them situated in the euphotic zone, spatially and temporally highly variable hypoxia has occurred, which has caused reductions of biomass and individual size (Figure 1.6). It may be more than a coincidence that some of the direct evidence for hypoxia-induced faunal reductions [Rosenberg and Loo, 1988; Josefson and Jensen, 1992a; Fallesen, 1992; Rosenberg et al., 1992] derives from the pycnocline depth in the Kattegat, where a large part of the primary production occurs [Richardson and Christoffersen, 1991] and where oxygenation immediately below this water layer may be restricted. However, only limited parts of the area seem to have reached the stage, where opportunistic polychaetes (e.g. *Capitella* spp.) dominate, or even the azoic stage.

References

Ærtebjerg, G., L. A. Jørgensen, P. Sandbeck, J. N. Jensen and H. Kaas, Vandmiljøplanens overvågningsprogram 1989 – Marine områder – Fjorde, kyster og åbent hav. *Faglig rapport fra DMU*, 8, 100 pp., 1990.

Alcaraz, M. and F. Sarda, Oxygen consumption by *Nephrops norvegicus* (L.) (Crustacea, Decapoda) in relationship with its moulting stage, *J. Exp. Mar. Biol. Ecol.*, 54, 113–118, 1981.

Andersin, A.-B., H. Cederwall, F. Gosselck, J. N. Jensen, A. B. Josefson, G. Lagzdins, H. Rumohr, and J. Warzocha, Zoobenthos, in *Baltic Sea Environmental Proceedings 35B, Second periodic assesment of the state of the marine environment of the Baltic Sea, 1984–1988, Background document*, pp. 211–275, 1990.

Atkinson, R. J. A., and A. C. Taylor, Physiological ecology of burrowing decapods, *Symp. zool. Soc. Lond.*, 59, 201–226, 1988.

Austen, M. C., J. B. Buchanan, H. G. Hunt, A. B. Josefson, and M. A. Kendall, Comparison of long-term trends in benthic and pelagic communities of the North Sea, *J. Mar. Biol. Ass. U.K.*, 71, 179–190, 1991.

Baden, S. P., M. H. Depledge, and L. Hagerman, Glycogen depletion and altered copper and manganese handling in *Nephrops norvegicus* following starvation and exposure to hypoxia, *Mar. Ecol. Prog. Ser.*, 103, 65–72, 1994.

Baden, S. P., L.-O. Loo, L. Pihl, and R. Rosenberg, Effects of eutrophication on benthic communities including fish: Swedish west coast, *Ambio*, 19, 113–122, 1990a.

Baden, S. P., L. Pihl, and R. Rosenberg, Effects of oxygen depletion on the ecology, blood physiology and fishery of the Norway lobster *Nephrops norvegicus*, *Mar. Ecol. Prog. Ser.*, 67, 141–155, 1990b.

Bagge, O., and S. Munch-Petersen, Some possible factors governing the catchability of Norway lobster in the Kattegat, *Rapp. P.-v. Réun. Cons. Perm. Intl Explor. Mer*, 175, 143–146, 1979.

Bagge, O., S. Mellergaard, and E. Nielsen, The importance of the benthos for bottom-living fish in southern Kattegat, *Havforskning fra Miljøstyrelsen*, 27, 106 pp., Danish Environmental Protection Agency, Copenhagen, 1993.

Bagge, O., E. Nielsen, S. Mellergaard, and J. Dalsgaard, Hypoxia and the demersal fish stock in the Kattegat (IIIa) and Subdivision 22, *ICES CM 1990 /E:4*, pp. 1–52, 1990.

Beukema, J. J., Changes in composition of bottom fauna of a tidal-flat area during a period of eutrophication, *Mar. Biol.*, 111, 293–301, 1991.

Beukema, J. J., and G. C. Cadée, Zoobenthos responses to eutrophication of the Dutch Wadden Sea, *Ophelia*, 26, 55–64, 1986.

Beukema, J. J., and G. C. Cadée, Growth rates of the bivalve *Macoma baltica* in the Wadden Sea during a period of eutrophication: relationships with concentrations of pelagic diatoms and flagellates, *Mar. Ecol. Prog. Ser.*, 68, 249–256, 1991.

Blegvad, G., Kvantitative undersøgelser af bundinvertebraterne i Kattegat med særlig henblik paa de for rødspætten vigtigste næringsdyr, *Beret. Danske Biol. Stn.*, 36, 1–55, 1930.

Brey, T., Increase in macrozoobenthos above the halocline in Kiel Bay comparing the 1960s with the 1980s, *Mar. Ecol. Prog. Ser.*, 28, 299–302, 1986.

Bridges, C. R., A comparative study of the respiratory properties and physiological function of haemocyanin in two burrowing and two non-burrowing crustaceans, *Comp. Biochem. Physiol.*, 83A, 261–270, 1986.

Bridges, C. R., and A. R. Brand, Oxygen consumption and oxygen independence in marine crustaceans, *Mar. Ecol. Prog. Ser.*, 2, 133–141, 1980.

Buchanan, J. B., Evidence of benthic pelagic coupling at a station off the Northumberland coast, *J. Exp. Mar. Biol. Ecol.*, 172, 1–10, 1993.

Cederwall, H., and R. Elmgren, Biomass increase of benthic macrofauna demonstrates eutrophication of the Baltic Sea, *Ophelia*, Supplement 1, 287–304, 1980.

Chapman, C. J., Ecology of juvenile and adult *Nephrops*, in *The Biology and Management of Lobsters*, edited by J. S. Cobbs and B. F. Phillips, Vol. 2, pp. 143–178, Academic Press, London, 1980.

Christensen, A. M., Feeding biology of the sea-star *Astropecten irregularis* Pennant, *Ophelia*, 8, 1–134, 1970.

Christensen, H., and E. Kanneworff, Sedimenting phytoplankton as major food source for suspension and deposit feeders in the Oresund, *Ophelia*, 24, 223–244, 1985.

Dauer, D. M., and W. G. Conner, Effects of moderate sewage input on benthic polychaete populations, *Estuar. Coast. Mar. Sci.*, 10, 335–346, 1980.

Detlefsen, V., and H. von Westerhagen, Oxygen deficiency and effects on bottom fauna in the eastern German Bight 1982, *Meeresforsch.*, 30, 42–53, 1983.

Depledge, M. H., and P. Bjerregaard, Haemolymph protein composition and copper levels in decapod crustaceans, *Helgoländer Meeresunters.*, 43, 207–223, 1989.

de Zwaan, A., and V. Putzer, Metabolic adaptions of intertidal invertebrates to environmental hypoxia: A comparison of environmental anoxia to exercise anoxia, in *Physiological Adaptions of Marine Animals*, edited by M. S. Laverack, pp. 33–62, *Symp. Soc. Exp. Biol.*, 39, 1985.

Dries, R.-R., and H. Theede, Sauerstoffmangelresistenz mariner Bodenevertebraten aus der Westlichen Ostsee, *Mar. Biol.*, 25, 327–333, 1974.

Dybern, B. I., and T. Høisæter, The burrows of *Nephrops norvegicus* (L.), *Sarsia*, 21, 49–55, 1965.

Evans, F., and A. Edwards, Changes in the zooplankton community off the coast of Northumberland between 1969 and 1988, with notes on changes in the phytoplankton and benthos, *J. Exp. Mar. Biol. Ecol.*, 172, 11–29, 1993.

Fallesen, G., How sewage discharge, terrestrial run-off and oxygen deficiencies affect the bottom fauna in Århus Bay, Denmark, in *Marine Eutrophication and Population Dynamics*, edited by G. Colombo, I. Ferrari, V. U. Ceccherelli, and R. Rossi, pp. 29–33, Olsen & Olsen, Fredensborg, Denmark, 1992.

Fallesen, G., and H. M. Jørgensen, Distribution of *Nephtys hombergii* and *N. ciliata* (Polychaeta: Nephtyidae) in Århus Bay, Denmark, with emphasis on the effect of severe oxygen deficiency, *Ophelia*, Supplement 5, 443–450, 1991.

Figureido, M. J., and H. J. Thomas, *Nephrops norvegicus* (Linnaeus 1758) Leach. – A review, *Oceanogr. Mar. Biol. Ann. Rev.*, 5, 371–407, 1967.

Garringa, L. J. A., Mobility of Cu, Cd, Ni, Pb, Zn, Fe and Mn in marine sediment slurries under anaerobic conditions and at 20% air saturation, *Neth. J. Sea Res.*, 27, 145–156, 1991.

Gaston, G. R., Effects of hypoxia on macrobenthos of the inner shelf off Cameron, Louisiana, *Estuar. coast. Shelf Sci.*, 20, 603–613, 1985.

Graf, G., W. Bengtsson, U. Diesner, R. Schulz, and H. Theede, Benthic response to sedimentation of a spring phytoplankton bloom: process and budget, *Mar. Biol.*, 67, 201–208, 1982.

Graham, M., Effects of trawling on animals of the sea bed, *Deep-Sea Res.*, 3, Supplement, 1–6, 1955.

Gray, J. S., Eutrophication in the sea, in *Marine Eutrophication and Population Dynamics*, edited by G. Colombo, I. Ferrari, V. U. Ceccherelli, and R. Rossi, pp. 3–15, Olsen & Olsen, Fredensborg, Denmark, 1992.

Gäde, G., Energy metabolism of arthropods and molluscs during environmental and functional anaerobiosis, *J. Exp. Zool.*, 228, 415–429, 1983.

Hagerman, L., Haemocyanin concentration in juvenile lobsters (*Homarus gammarus*) in relation to moulting cycle and feeding condition, *Mar. Biol.*, 77, 11–17, 1983.

Hagerman, L., Haemocyanin concentration in the shrimp *Crangon crangon* (L.) after exposure to moderate hypoxia, *Comp. Biochem. Physiol.*, 85A, 721–724, 1986.

Hagerman, L., and S. P. Baden, *Nephrops norvegicus*: field study of effects of oxygen deficiency on haemocyanin concentration, *J. Exp. Mar. Biol. Ecol.*, 116, 135–142, 1988.

Hagerman, L., and A. Szaniawska, Anaerobic metabolic strategy of the glacial relict isopod *Saduria (Mesidothea) entomon, Mar. Ecol. Prog. Ser.*, 59, 91–96, 1990.

Hagerman, L., and A. Szaniawska, *Saduria entomon*, ecophysiological adaptations for survival in the Baltic, in *Proc. 12th Baltic Marine Biologists Symp.*, edited by E. Bjørnestad, L. Hagerman, and K. Jensen, pp. 71–76, Olsen & Olsen, Fredensborg, Denmark, 1992.

Hagerman, L., and R. F. Uglow, Effects of hypoxia on the respiratory and circulatory regulation of *Nephrops norvegicus, Mar. Biol.*, 87, 273–278, 1985.

Hagerman, L., and B. Vismann, Anaerobic metabolism, hypoxia and hydrogen sulphide in the brackish water isopod *Saduria entomon, Ophelia*, 38, 1–11, 1993.

Hagerman, L., T. Søndergaard, K. Weile, D. Hosie, and R. F. Uglow, Aspects of blood physiology and ammonia excretion in *Nephrops norvegicus* under hypoxia, *Comp. Biochem. Physiol.*, 97A, 51–55, 1990.

Hallbäck, H., and M. Ulmestrand, Havskräfta i Kattegatt, *Fauna och Flora*, 85, 186–192, 1990.

Huettel, M., and G. Gust, Impact of bioroughness on interfacial solute exchange in permeable sediments, *Mar. Ecol. Prog. Ser.*, 89, 253–267, 1994.

ICES, *ICES Doc. CM 1992 / Assess: 12*, 1992.

ICES, *ICES Doc. CM 1993 / Assess: 16*, 1993.

Johnson, R. K., and T. Wiederholm, Pelagic-benthic coupling – The importance of diatom interannual variability for population oscillations of *Monoporeia affinis, Limnol. Oceanogr.*, 37, 1596–1607, 1992.

Josefson, A. B., Increase in benthic biomass in the Skagerrak–Kattegat during the 1970s and 1980s – effects of organic enrichment?, *Mar. Ecol. Prog. Ser.*, 66, 117–130, 1990.

Josefson, A. B., and B. Widbom, Differential response of benthic macrofauna and meiofauna to hypoxia in the Gullmar Fjord basin, *Mar. Biol.*, 100, 31–40, 1988.

Josefson, A. B., and J. N. Jensen, Effects of hypoxia on soft-sediment macrobenthos in southern Kattegat, in *Marine Eutrophication and Population Dynamics*, edited by G. Colombo, I. Ferrari, V. U. Ceccherelli, and R. Rossi, pp. 21–28, Olsen & Olsen, Fredensborg, Denmark, 1992a.

Josefson, A. B., and J. N. Jensen, Growth patterns of *Amphiura filiformis* support the hypothesis of organic enrichment in the Skagerrak–Kattegat area, *Mar. Biol.*, 112, 615–624, 1992b.

Josefson, A. B., J. N. Jensen, and G. Ærtebjerg, The benthos community structure anomaly in the late 1970's and early 1980's – a result of a major food pulse?, *J. Exp. Mar. Biol. Ecol.*, 172, 31–45, 1993.

Jørgensen, B. B., Seasonal oxygen depletion in the bottom waters of a Danish fjord and its effect on the benthic community, *Oikos*, 34, 68–76, 1980.

Kröncke, I., Macrofauna standing stock of the Dogger Bank. A comparison: III. 1950–54 versus 1985–87. A final summary, *Helgoländer Meeresunters.*, 46, 137–169, 1992.

Kwee, D. J., Ammonia effluxes in association with cardiac and ventilatory activities in *Nephrops norvegicus* (L.), *Necora puber* (L.) and *Cancer pagurus* (L.). Ph.D. thesis, Dept. of Appl. Biology, University of Hull, Hull, 1993.

Lallier, F., F. Boitel, and J. P. Truchot, The effect of ambient oxygen and temperature on haemolymph l-lactate and urate concentrations in the shore crab *Carcinus maenas, Comp. Biochem. Physiol.*, 86A, 255–260, 1987.

Leonhard, S. and S. Varming, Bundfauna i Lillebælt 1911–1990, *Lillebæltsamarbejdet*, 185 pp., 1992

Levin, L. A., Effects of enrichment on reproduction in the opportunistic polychaete *Streblospio benedicti* (Webster): a mesocosm study, *Biol. Bull. Mar. Biol. Lab. Woods Hole*, 171, 143–160, 1986.

Mangum, C. P., Evaluation of the functional properties of invertebrate hemoglobins, *Neth. J. Sea Res.*, 7, 303–315, 1973.

Mangum, C. P., Oxygen transport in the blood, in *The Biology of Crustacea, Vol. 5. Internal Anatomy and Physiological Regulation*, edited by L. H. Mantel, pp. 373–430, Academic Press, New York, 1983.

Niermann, U., E. Bauerfeind, W. Hickel, and H. V. Westernhagen, The recovery of benthos following the impact of low oxygen content in the German Bight, *Neth. J. Sea. Res.*, 25, 215–226, 1990.

Oeschger, R., Long-term anaerobiosis in sublittoral marine invertebrates from the Western Baltic Sea: *Halicryptus spinolosus* (Priapulida), *Astarte borealis* and *Arctica islandica* (Bivalvia), *Mar. Ecol. Prog. Ser.*, 59, 133–143, 1990.

Pearson, T. H., and R. Rosenberg, Macrobenthic succession in relation to organic enrichment and pollution of the marine environment, *Oceanogr. Mar. Biol. Ann. Rev.*, 16, 229–311, 1978.

Pearson, T. H., A. B. Josefson, and R. Rosenberg, Petersen's benthic stations revisited. I. Is the Kattegatt becoming eutrophic?, *J. Exp. Mar. Biol. Ecol.*, 92, 157–206, 1985.

Petersen, C. G. J., Havets Bonitering. II. Om Havbundens Dyresamfund og om disses betydning for den marine Zoogeografi, *Beret. Minist. Landbr. Fisk. Dan. Biol. Stn*, 21, 1–42, 1913.

Petersen, C. G. J., Om Havbundens dyresamfund i Skagerak, Kristiamiafjord og de danske Farvande, *Beret. Minist. Landbr. Fisk. Dan. Biol. Stn*, 23, 5–26, 1915.

Petersen, J. K., and L. Pihl, Responses to hypoxia of plaice (*Pleuronectes platessa*) and dab (*Limanda limanda*) in the South-East Kattegat: Distribution and growth, *Envir. Biol. Fish.*, 43, 311–321, 1995.

Peterson, C. H., Predation, competitive exclusion, and diversity in the soft-sediment benthic communities of estuaries and lagoons, in *Ecological Processes in Coastal and Marine Ecosystems*, edited by R.L. Livingston, pp. 233–264, Plenum Press, New York, 1979.

Pihl, L., Effects of oxygen depletion on demersal fish in coastal areas of the south-east Kattegat, in *Reproduction, Genetics and Distributions of Marine Organisms*, edited by J. S. Ryland, and P. A. Tyler, pp. 431–439, Proc. 23th European Marine Biology Symp., Olsen & Olsen, Fredensborg, Denmark, 1989.

Rhoads, D. C., and L. F. Boyer, The effects of marine benthos on physical properties of sediments: a successional perspective, in *Animal-Sediment Relations*, edited by P. L. McCall, and M. Tevesz, pp. 3–52, Plenum Press, New York, 1982.

Rice, A. L., and J. Chapman, Observations on the burrows and burrowing behaviour of two mud-dwelling decapod crustaceans, *Nephrops norvegicus* and *Goneplax rhomboides*, *Mar. Biol.*, 10, 330–342, 1971.

Richardson, K., and A. Christoffersen, Seasonal distribution and production of phytoplankton in the southern Kattegat, *Mar. Ecol. Prog. Ser.*, 78, 217–227, 1991.

Richardson, K., and J. Heilmann, Primary production in the Kattegat: past and present, *Ophelia*, 41, 317–328, 1995.

Rosenberg, R., and L.-O. Loo, Marine eutrophication induced oxygen deficiency: effects on soft bottom fauna, western Sweden, *Ophelia*, 29, 213–225, 1988.

Rosenberg, R., L.-O. Loo, and P. Möller, Hypoxia, salinity and temperature as structuring factors for marine benthic communities in a eutrophic area, *Neth. J. Sea. Res.*, 30, 121–129, 1992.

Rosenberg, R., and P. Möller, Salinity stratified benthic macrofaunal communities and long-term monitoring along the west coast of Sweden, *J. Exp. Mar. Biol. Ecol.*, *37*, 175–203, 1979.

Rosenberg, R., J. S. Gray, A. B. Josefson, and T. H. Pearson, Petersen's benthic stations revisited. II. Is the Oslofjord and eastern Skagerrak enriched?, *J. Exp. Mar. Biol. Ecol.*, *105*, 219–251, 1987.

Rosenberg, R., B. Hellman, and B. Johansson, Hypoxic tolerance of marine benthic fauna, *Mar. Ecol. Prog. Ser.*, *79*, 127–131, 1991.

Sandberg, E., Does short-term oxygen depletion affect predator-prey relationships in zoobenthos?. Experiments with the isopod *Saduria entomon*, *Mar. Ecol. Prog. Ser.*, *103*, 73–80, 1994.

Schlütter, H., Haemocyanins iltbindingsevne og affinitetsmodulerende faktorer hos jomfruhummeren, *Nephrops norvegicus*, *Rep. Mar. Biol. Lab., Helsingør*, pp. 1–63, 1993.

Schöttler, U., An investigation on the anaerobic metabolism of *Nephtys hombergii* (Annelida: Polychaeta), *Mar. Biol.*, *71*, 265–269, 1982.

Schöttler, U., B. Surholt, and E. Zebe, Anaerobic metabolism in *Arenicola marina* and *Nereis diversicolor* during low tide, *Mar. Biol.*, *81*, 69–73, 1984.

Schöttler, U., D. Daniels, and K. Zapf, Influence of anoxia on adaptation of euryhaline polychaetes to hyposmotic conditions, *Mar. Biol.*, *104*, 443–451, 1990.

Suess, E., Particulate organic carbon flux in the oceans-surface productivity and oxygen utilization, *Nature* (Lond.), *288*, 260–263, 1980.

Temara, A., C. De Ridder, J. G. Kuenen, and L. A. Robertson, Sulfide-oxidizing bacteria in the burrowing echinoid *Echinocardium cordatum* (Echinodermata), *Mar. Biol.*, *115*, 179–185, 1993.

Thamdrup, B., R. N. Glud, and J. W. Hansen, Manganese oxidation and in situ manganese fluxes from a coastal sediment, *Geochim. Cosmochim. Acta*, *58*, 2563–2570, 1994.

Theede, H., Physiological approaches to environmental problems of the Baltic, *Limnologica (Berl.)*, *15*, 443–458, 1984.

Theede, H., A. Ponat, K. Hiroki, and C. Schlieper, Studies on the resistance of marine bottom invertebrates to oxygen-deficiency an hydrogen sulphide, *Mar. Biol.*, *2*, 325–337, 1969.

Truchot, J. P., Lactate increases the oxygen affinity of crab hemocyanin, *J. Exp. Zool.*, *214*, 205–208, 1980.

Truchot, J. P., Regulation of acid base balance, in *The Physiology of Crustacea. Vol. 5. Internal Anatomy and Physiological Regulation*, edited by L. H. Mantel, pp. 431–457, Academic Press, New York, 1983.

Uglow, R. F., Haemolymph protein concentrations in portunid crabs. II. The effect of imposed fasting on *Carcinus maenas*, *Comp. Biochem. Physiol.*, *31*, 959–967, 1969.

Vismann, B., Sulfide detoxification and tolerance in *Nereis (Hediste) diversicolor* and *Nereis (Neanthes) virens* (Annelida: Polychaeta), *Mar. Ecol. Prog. Ser.*, *59*, 229–238, 1990.

Vismann, B., Sulfide tolerance: Physiological mechanisms and ecological implications, *Ophelia*, *34*, 1–27, 1991.

Vismann, B., Physiology of sulfide detoxification in the isopod *Saduria (Mesidotea) entomon*, *Mar. Ecol. Prog. Ser.*, *76*, 283–293, 1991.

Weigelt, M., Short- and long-term changes in the benthic community of the deeper parts of Kiel Bay (Western Baltic) due to oxygen depletion and eutrophication, *Meeresforsch.*, *33*, 197–224, 1991.

Weigelt, M., and H. Rumohr, Effects of wide-range oxygen depletion on benthic fauna and demersal fish in Kiel Bay 1981–1983, *Meeresforsch.*, *31*, 124–136, 1986.

Westernhagen, H. V., W. Hickel, E. Bauerfeind, U. Niermann, and I. Kröncke, Sources and effects of oxygen deficiencies in the south-eastern North Sea, *Ophelia*, *26*, 457–473, 1986.

Zeis, B., A. Nies, C. R. Bridges, and M. K. Grieshaber, Allosteric modulation of haemocyanin oxygen-affinity by l-lactate and urate in the lobster *Homarus vulgaris*. I. Specific and additive effects on haemocyanin oxygen-affinity, *J. Exp. Biol.*, *168*, 93–110, 1992.

9

Shallow Waters and Land/Sea Boundaries

Jens Borum

9.1. Introduction

Two important issues related to eutrophication and shallow-water ecology will be discussed in this chapter. The first concerns the role of shallow waters as recipients, transporters and sinks of nutrient loading from land. Attention is primarily directed toward nitrogen, as it is this nutrient that is believed to be the main limiting nutrient for plant growth and primary production in temperate coastal waters [Ryther and Dunstan, 1971; Granéli et al., 1986]. The second issue addressed is how nutrient enrichment influences the autotrophic communities and the primary production on which all other biotic components and processes within the coastal zone ultimately depend.

For practical purposes, "shallow" coastal waters in the present context are defined as marine and brackish areas with water depths of less than 20 meters which represents the approximate depth limit to which benthic plants may contribute substantially to primary production. This zone includes a large part of the marine Danish areas, extending from a few to more than 50 kilometers from the coast, and covers an area approximately of the same size as the land area. The topography of Denmark is of glacial origin and estuaries and embayments are typically shallow with sandy or silty sediments. Although the Danish term "fjord" is a common name for estuarine areas, it should not be confused with the classical Norwegian fjord. Only a few Danish fjords have a sill and deep inner basins. The tidal amplitude varies from 2 meters in the Wadden Sea to a few centimeters in some estuaries and in parts of the Belt Sea.

9.2. Nutrient runoff

Reliable estimates of total nutrient runoff from land only exist from the mid-1970s (Figure 9.1) and, therefore, do not include the period of exponential increase in agricultural and urban nutrient runoff during the 1950s and 1960s. Point source

Eutrophication in Coastal Marine Ecosystems
Coastal and Estuarine Studies, Volume 52, Pages 179–203
Copyright 1996 by the American Geophysical Union

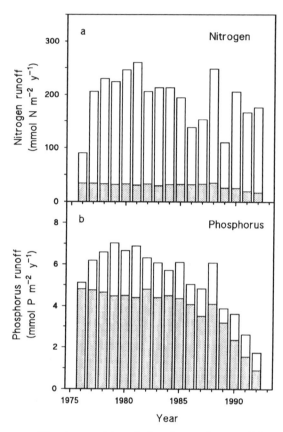

Figure 9.1. Area specific nitrogen (a) and phosphorus (b) runoff from the county of Fyn (3505 km²), Denmark. Contributions from urban sewage, industrial waste water and other point sources (shaded bars) and from diffuse sources (open bars). Data from Fyns Amt [1993].

nitrogen contributions were relatively constant from 1976 to 1988 and thereafter declined due to the implementation of advanced sewage treatment (Figure 9.1a). The largest contribution to nitrogen runoff, however, is from diffuse sources, which remain highly variable from year to year depending on climate and rainfall. Consistent with a constant application of fertilizers since 1980, no systematic reduction of nitrogen runoff has been observed from 1976 to 1992 (Figure 9.1a).

The phosphorus runoff was slowly reduced from 1976 to 1988 but after 1988 declined rapidly and is now only 20–25% of the level before 1988 (Figure 9.1b). Consequently, Danish coastal areas have experienced a substantial reduction in phosphorus loading, while nitrogen loading has been more or less constant during the last 15–20 years. Typically, nitrogen loading per square meter of Danish coastal waters is of the same magnitude as, or even larger than, nutrient application to heavily fertilized fields.

9.3. Nutrient Retention in Coastal Ecosystems

The coastal zone plays an important role as a filter inserted between land and ocean by retaining suspended materials and nutrients from land [Sharp et al., 1984; Nixon and Pilson, 1984]. Basic understanding and quantitative knowledge of nutrient retention and losses in coastal waters are essential to evaluate and predict open-water eutrophication. One needs to know how much of land runoff actually reaches the open waters to be able to evaluate the present role of runoff versus nutrient deposition from the atmosphere (Figure 9.2).

Nitrogen and phosphorus from land enter the biogeochemical cycles of the coastal areas and are lost by sediment burial and, in the case of nitrogen, by bacterial denitrification under anoxic conditions (Figure 9.2). Nutrients contained in imported or autochtonously produced organic particles settle on the sediment surface or are subject to biodeposition by benthic suspension feeders. Rates of gross sedimentation in shallow marine areas are very high due to frequent resuspension of the sediment surface layers (cf. chapters 6 and 7), and the traditional use of sediment traps gives poor information on net sedimentation rates. Net rates can be measured by analysis of sediment depth profiles of nutrients combined with ^{210}Pb dating, but time resolution is poor, in the order of decades, due to resuspension and bioturbation of the estuarine sediments [Floderus, 1992]. Therefore, we have no simple

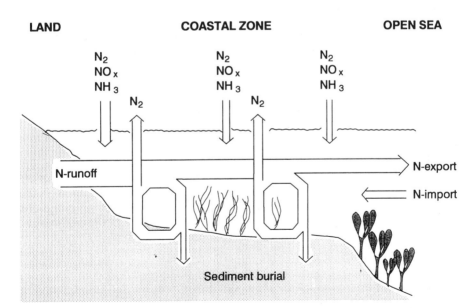

Figure 9.2. Schematic presentation of nitrogen loading, retention and transport within the coastal zone. Nutrients from land are gradually retained by sediment burial or lost through denitrification during transport from land to open sea, while the relative importance of atmospheric and oceanic input increases with distance from land.

TABLE 1. Temperate coastal ecosystems for which nutrient mass balances have been established. Localities, mean nutrient inputs, and references.

Locality	mmol m^{-2} d^{-1} N input	P input	Sources
Narragansett Bay	6.6	–	Seitzinger et al., 1984
Lough Hyne	5.4	0.31	M. Johnson, pers. comm.
Bothnian Bay	0.54	0.006	Wulff et al., 1990
Baltic Sea	0.65	0.016	Wulff et al., 1990
Roskilde Fjord area 1	6.7	0.41	Kamp-Nielsen, 1992
" area 2	6.2	0.34	Kamp-Nielsen, 1992
" area 3	6.9	0.29	Kamp-Nielsen, 1992
" area 4	5.9	0.25	Kamp-Nielsen, 1992
Limfjorden	4.8	–	County of Viborg
Ringkøbing Fjord (1987–1991)	4.6	0.10	County of Ringkøbing
Mariager Fjord (1986–1992)	6.4	–	County of Nordjylland
Norsminde Fjord	23.9	–	County of Aarhus

means of resolving seasonal or annual changes in sediment nutrient burial following alteration in anthropogenic nutrient loading.

While benthic burial is the main phosphorus sink during transport from land to open sea, sediment accumulation of nitrogen accounts for less than 20% of total nitrogen removal in estuaries. The rest is lost by reduction of nitrate to gaseous nitrogen [Seitzinger, 1988; 1990; Billen et al., 1991; chapter 6, this vol.]. The denitrification process is driven by nitrate diffusing into the sediment from the water column and by coupled nitrification-denitrification within the sediment [Jenkins and Kemp, 1984]. Denitrification rates in coastal marine areas range from 0 to 500 μmol N m^{-2} h^{-1} [Seitzinger, 1988], exhibiting substantial local variability but also showing the potential importance of denitrification as a nitrogen sink. Despite the large spatial and temporal variability of measured denitrification rates, estimates of annual denitrification in very diverse Danish coastal areas exhibited surprisingly moderate variation around a mean value of 400 μmol N m^{-2} d^{-1}, corresponding to 20 kg N ha^{-1} yr^{-1} [Nielsen et al., 1994].

Nutrient mass balances established for whole marine ecosystems, as those known from fresh waters [e.g., Ahlgren, 1967; Andersen, 1971], are subject to considerable error because of the dynamic water exchange with adjacent seas and few figures have been published [e.g., Wulff et al., 1990; Kamp-Nielsen, 1992]. Reported mass balances and a number of unpublished Danish mass balance estimates for nutrients in estuaries (Table 1) suggest that steady-state phosphorus retention (Figure 9.3a) ranges from a few to almost 80% of total phosphorus input (input = loading from land + import from adjacent seas, where import is estimated based on the annual salt balance of the coastal area and annual means of total nutrient concentrations in the water imported from adjacent seas). Phosphorus retention is usually less than 20% of the input (Figure 9.3a) [Billen et al., 1991], but in coastal areas with long residence times, such as the Baltic Sea, phosphorus removal may exceed 50%

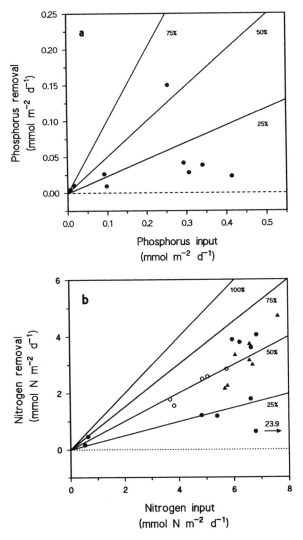

Figure 9.3. Phosphorus (a) and nitrogen (b) retention versus total phosphorus and nitrogen input to the coastal areas listed in Table 1. Only coastal areas presumably in steady state are included. ○: Ringkøbing Fjord, different years; ▲: Mariager Fjord, different years; ●: other areas.

[Wulff et al., 1990]. Nitrogen retention (permanent burial + denitrification) ranges from 2 to 65% of nitrogen input (input = loading from land + nitrogen fixation + atmospheric deposition + import from adjacent seas; Figure 9.3b). Absolute nitrogen retention rates increase with increasing nitrogen loading and vary between 0.25 and 5 mmol N m^{-2} d^{-1}, a range comparable to published rates of directly measured denitrification from estuarine areas [Hattori, 1983; Seitzinger, 1988].

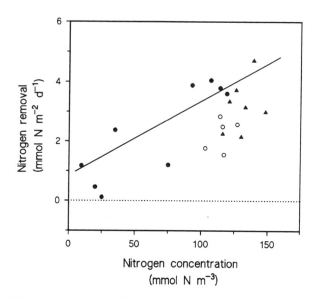

Figure 9.4. Nitrogen retention (sediment burial plus denitrification) versus average annual nitrogen concentration in coastal areas listed in Table 1. Mean values for Ringkøbing Fjord (○) and Mariager Fjord (▲) are included in the described relationship ($r^2 = 0.54$, n = 11).

Seitzinger [1988, 1990] reported that approximately 50% of the inorganic nitrogen loading from land is lost through denitrification independent of loading rates and of hydraulic and bathymetric differences among estuaries. This observation has tended to translate into a general "50% of total nitrogen loading"-rule in the literature. Apart from being highly unlikely due to the marked differences in bathymetry and flushing, the 50%-rule predicts that changes in nitrogen loading do not influence percentage losses despite major concurrent changes in the biological structure of the shallow coastal areas. The present compilation of data falsifies the 50%-rule (Figure 9.3b). Nitrogen retention ranges from 2 to 75% of total nitrogen import among different coastal areas. In addition, tidal coastal areas with short water residence times have negligible percentage losses [Pejrup et al., 1993].

It is to be expected that areas subject to very high rates of nitrogen loading and/or with short water residence times must have low percentage nitrogen retention. In order to be denitrified or permanently buried, imported nitrogen needs to be assimilated and transported down to the sediment. In areas of short residence times, only a small fraction of the imported nitrogen may reach the sediment and, therefore, losses relative to imports are low. In slowly flushed coastal areas, imported nitrogen may, however, pass the sediment compartment several times, and relative losses exceed the 50% [Wulff et al., 1990]. To account for differences in flushing, nitrogen losses might be better estimated from average water column concentrations of total nitrogen (Figure 9.4), However, substantial residual variation still remains unexplained and likely reflects differences in bathymetry, physical mixing patterns and biological structure.

9.4. Response to Reduced Nutrient Loading

We have no broad empirical basis to predict how nutrient losses of coastal ecosystems respond to reduced loading. As emphasized later, benthic plant populations, and probably also the benthic infauna, will benefit from reduced nutrient loading and their influence on coastal nitrogen cycling may affect the proportions of nitrogen retention due to changes in denitrification [Caffrey and Kemp 1990; Kristensen, 1988; Pelegrí et al., 1994]. Benthic infauna bioturbates and many invertebrate species construct and irrigate burrows. During ventilation, oxygen and nitrate from the water column are transported to the burrow lumen and stimulate mineralization, nitrification and denitrification within the sediment (chapter 6). Stimulated nitrification and denitrification are primarily due to the areal increase in the anoxic-oxic transition zone and burrow dimensions and density are, therefore, key factors determining the effects of infauna on nitrogen cycling [Aller, 1988; Pelegrí et al., 1994]. If, indeed, the density of burrowing invertebrates increases with reduced nutrient loading, nitrogen losses through denitrification should be enhanced due to direct reduction of nitrate from the water column or to an increase of coupled nitrification-denitrification. Conversely, faunal stimulation of sediment mineralization and subsequent release of NH_4^+ and urea to the water column may reduce sediment burial of nitrogen and, hence, may just shift the balance between the two nitrogen sinks.

Stimulation of benthic microalgae and macrophytes due to improved light conditions following lower nutrient loading also affects the dynamics of the anoxic-oxic transition zone in marine sediments [Revsbech et al., 1981]. During biomass build-up, benthic microalgae provide a net oxygen input to the sediment, and induce diel changes in oxygen concentrations leading to rhythmic changes in the depth position of the anoxic-oxic interface and in denitrification rates [Revsbech et al., 1981; Risgaard-Petersen et al., 1994]. The net effect of stimulated growth of benthic microalgae, however, seems to be a marked inhibition of denitrification [Nielsen et al., 1994]. Oxygen release from roots of submerged macrophytes during photosynthesis stimulates both nitrification and denitrification [Caffrey and Kemp, 1990], and, therefore, increased abundance of rooted macrophytes should enhance nitrogen losses.

With the possible exception of benthic microalgae, the integrated effect of biotic components on denitrification suggests that relative nitrogen losses should increase with reduced loading. This prediction is, however, contradictory to the more general viewpoint that nutrient utilization and conservation are more efficient in mature, nutrient-poor ecosystems than in perturbed, nutrient-rich ecosystems [Odum, 1971]. Though the detailed studies on regulation of denitrification suggest an increased loss of available nitrogen with higher abundance of benthic fauna and rooted macrophytes, the net result may well turn out to be the opposite at the ecosystem level, because inorganic nitrogen availability is reduced as a response to reduced loading. For example, extended depth penetration of rooted macrophytes may improve sediment oxygen conditions but a larger proportion of the total nitrogen pool within the ecosystem may also be bound within plant biomass and consequently be less available to denitrifying bacteria.

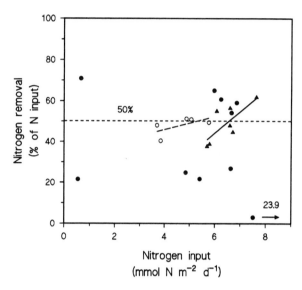

Figure 9.5. Percentage nitrogen retention versus nitrogen input to the coastal areas listed in Table 1. Relationships for different years in Ringkøbing Fjord (dashed, r^2 = 0.389, p = 0.26) and Mariager Fjord (solid, r^2 = 0.579, p = 0.047) are shown.

There does not appear to be a clear relationship between the proportional loss of nitrogen and the nitrogen input among different coastal ecosystem [Billen et al., 1991]. Year-to-year variation within two Danish estuaries suggests that the proportion of nitrogen retention declines with reduced nitrogen loading (Figure 9.5). However, since a systematic change in loading has not occurred over several years in these estuaries and since major changes in benthic flora and fauna often have response times longer than one year, the observed changes in relative nitrogen losses may well reflect transient responses and not tell us what happens on a longer time scale. Reduced nitrogen loading to a shallow Danish embayment, Kertinge Nor, caused this system to become net nitrogen exporting during the first two years after loading was reduced (chapter 10), but a new steady state has not yet been reached.

The general decline in phosphorus loading to Danish estuarine systems since 1988 has provided more information on systems responses to reduced P loading. However, as in the case of nitrogen dynamics in Kertinge Nor, new steady-state mass balances have not yet been established. In the large and very shallow (mean depth: 1.9 m) Ringkøbing Fjord, P loading was reduced by 50–60% from 1988 to 1991 [Ringkøbing Amtskommune, 1992]. Export of phosphorus from the fjord to the North Sea remained high, and the ecosystem changed from being a P-retaining to a net P-exporting system [Ringkøbing Amtskommune, 1992]. Though water residence time in Ringkøbing Fjord is short (0.3 yr) it is expected to take more years before a new steady-state P retention has been reached.

The shallow coastal zone does indeed function as an important nutrient filter and we have gained substantial knowledge on the mechanisms of process regulation and can also provide rough estimates of nutrient retention. Our ability to offer more accurate estimates of nutrient retention and predict changes in retention at the systems level, however, still needs to be improved. We have fair ideas on how, for example, rooted macrophytes influence rates of denitrification in the root zone and can also describe the basic influence of increased plant abundance on distribution and availability of nitrogen pools on a local scale. The overall effect of these complex relationships in terms of ecosystem nutrient retention can, however, hardly be determined using a mechanistic approach. Therefore, we need to put more emphasis on mass balance studies in coastal ecosystems of different loading, hydraulics and morphometry and, moreover, follow systems responses to changes in loading. Another attractive possibility is to assess nutrient dynamics in manipulated mesocosms. However, mesocosm experiments should be designed and operated so that they simulate realistic changes in the biological structure of both water column and benthic communities, i.e. they must be spatially and temporally scaled according to the sizes and generation times of the important benthic components and not only to those of the plankton.

9.5. Coastal Plant Communities and Eutrophication

Increased nutrient runoff has severe consequences for the ecology of coastal waters, even though the coastal zone is already eutrophic compared to many other aquatic environments before anthropogenic loading was initiated. Nutrient-enriched coastal waters exhibit periods of hypoxia or anoxia and mass mortality among benthic invertebrates and fish may occur. In the literature, these events are frequently explained as a result of increased production of organic matter within the ecosystem due to stimulation of the planktonic microalgae as observed in eutrophied, deep lakes and open marine waters. Aquatic ecologists often forget, however, that benthic plant communities are important contributors to the autotrophic production in pristine coastal areas and, therefore, neglect benthic plant responses to nutrient enrichment.

Benthic macrophytes attain dense populations which are as productive as some of the most productive terrestrial plant communities [Westlake, 1963; Mann, 1972; McRoy and McMillan, 1977]. Since eutrophication seems to inhibit rather than stimulate benthic plant populations [Cambridge et al., 1986; Sand-Jensen and Borum, 1991], nutrient enrichment of shallow coastal areas may not stimulate the total primary production per unit area but rather shift the main productivity from the benthic to the planktonic community. The fundamentals of eutrophication phenomena in coastal waters can only be understood by including the effects on all autotrophic components. Accordingly, the basic questions are, how and why different plant components respond to nutrient enrichment and how is the integrated response of total primary production to eutrophication?

The high productivity in coastal waters originates from a wide variety of plants ranging in size from picoplankton of less than 1 μm in diameter to 30 m long perennial macroalgae. Pristine shallow waters of temperate regions are colonized by rooted seagrasses and benthic microalgae on sediments and by perennial macroalgae on rocks and hard bottom sediments. If phytoplankton biomasses are moderate, light penetration through the water column is high and the primary production of benthic plants may exceed phytoplankton production substantially in these coastal ecosystems [Mann, 1972; Walker et al., 1988].

A characteristic change in the qualitative composition of plant communities is observed following increased nutrient inputs to embayments and estuaries [Sand-Jensen and Borum, 1991]. Phytoplankton biomasses increase [Boynton et al., 1982; Nixon and Pilson, 1983; Monbet, 1992], the abundance of epiphytic microalgae on seagrasses and on macroalgae increases [Borum, 1985; Twilley et al., 1985; Cambridge et al., 1986], and ephemeral macroalgae such as *Ulva lactuca* become more abundant and may accumulate in great masses [Harlin and Thorne-Miller, 1981; Valiela et al., 1992; Sfriso et al., 1992]. Concurrent with these changes, the depth penetration and abundance of seagrasses and perennial macroalgae decline [Borum, 1983; Kautsky et al., 1986; Cambridge et al., 1986]. Elements of this sequential development have been observed from many coastal areas and have also been experimentally documented using mesocosm manipulation [Twilley et al., 1985; Fong et al., 1993]. Accordingly, nutrient enrichment of shallow coastal waters changes the balance among autotrophic components from dominance of perennial macroalgae and seagrasses toward dominance of ephemeral macroalgae and pelagic microalgae.

9.6. Nutrient Requirements and Supplies

It is not very clear how the total nutrient concentration in the water column controls plant community composition. Microalgae and ephemeral macroalgae have large surface area to volume (SA:V) ratios, and are known to have higher rates of nutrient uptake per unit of biomass than large plants with low SA:V ratios (cf. chapter 4). With high rates of uptake per biomass, yet at limiting nutrient concentrations, small microalgae are usually assumed to be better able to meet their nutrient requirements for growth than plants with low SA:V ratios and, therefore, small microalgae should be competitively superior to larger in oligotrophic waters. This explanation seems intuitively plausible and is supported by observations of changes in the size composition of phytoplankton communities in open waters (cf. chapter 4). It does not explain, however, why large, slow-growing benthic plants with low SA:V ratios dominate in shallow coastal ecosystems with low nutrient concentrations.

Dominance of any given phototrophic organism implies that it is able to sustain a constant and sufficiently high biomass by compensating losses of nutrients and other resources through exploitation of these resources in the environment. If relative losses are low, as they are for larger plants where grazing losses are much lower than for microalgae [Duarte, 1995], gains can be low. Therefore, successful resource acquisition cannot be evaluated from absolute rates of gain alone (e.g. rates of nutrient uptake per biomass) but must be compared to the demand of the organism and its ability to satisfy this demand through various mechanisms [Sand-Jensen and Borum, 1991].

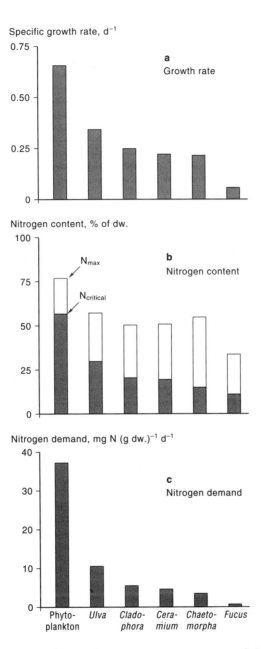

Figure 9.6. Maximum specific growth rates (a), nitrogen contents of tissue (b), and nitrogen demands needed to sustain maximum growth (c) of phytoplankton, *Ulva lactuca*, *Cladophora sericea*, *Ceramium rubrum*, *Chaetomorpha linum* and *Fucus vesiculosus* in Roskilde Fjord, Denmark. Nitrogen content is shown as critical content (shaded bars) and maximum observed content (shaded plus open bars). Data from Pedersen [1993].

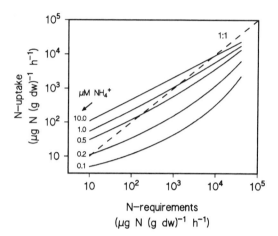

Figure 9.7. Relationships between ammonium uptake rates at different external ammonium concentrations (numbers) and nitrogen requirements during maximum growth of different marine algae ranging in size from planktonic microalgae to large kelps (M. Hein and M.F. Pedersen, unpublished data). Dashed line indicates balance between uptake and requirement.

The internal concentration of nutrients (e.g. total nitrogen content per plant biomass unit) required to saturate plant growth varies with the genetically fixed maximum growth rate of the plant. Small organisms typically have high growth rates and require high internal nutrient concentrations [Duarte, 1995; Smith and Kalff, 1982]. According to Redfield, the average C:N:P molar ratio for planktonic algae is 106:16:1 [Redfield et al., 1963]. The slow-growing macroalgae have much higher C:N:P ratios (ca. 800:49:1 [Duarte, 1992]) but some fast-growing macroalgal species with simple morphology have ratios closer to that of phytoplankton (e.g. *Porphyra* sp.: 137:23:1 [Atkinson and Smith, 1984]). Seagrasses are slow-growing and have an average C:N:P ratio of 474:24:1 [Duarte, 1990].

Pedersen [1993] examined and compared nitrogen acquisition and growth of different types of marine algae in a shallow coastal area. The maximum, specific growth rate ranged from 0.065 d^{-1} in the perennial brown alga, *Fucus vesiculosus*, to 0.66 d^{-1} for a natural phytoplankton community, while growth rates of four ephemeral macroalgae were between these extremes (Figure 9.6). The critical nitrogen content (sensu Hanisak [1979]), below which algal growth is nutrient- limited, varied five-fold and, together with the differences in growth rates, resulted in a 50-fold higher nitrogen demand per unit biomass and time for phytoplankton than for *Fucus vesiculosus* (Figure 9.6). Thus, small plants require a much higher nutrient supply rate than large, slow-growing plants to sustain maximum growth.

The biomass-specific nutrient uptake of microalgae is much faster than that of, for example, seagrasses and kelps [e.g., Duarte, 1995], although an analysis of the relationship between maximum inorganic nitrogen uptake and SA:V ratios showed that perennial macroalgae with low SA:V actually have the same or even larger uptake capacity per unit of surface area compared to microalgae [Hein et al., 1995]. Nutri-

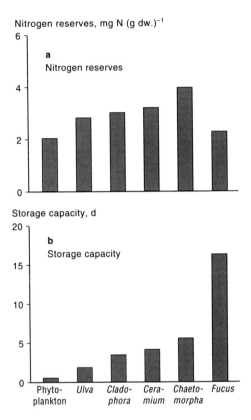

Figure 9.8. Nitrogen reserves ($N_{max.}$ minus $N_{critical}$; a) and nitrogen storage capacity (the time nitrogen reserves can sustain maximum growth; b) of phytoplankton, *Ulva lactuca, Clado- phora sericea, Ceramium rubrum, Chaetomorpha linum* and *Fucus vesiculosus* in Roskilde Fjord, Denmark. Data from Pedersen [1993].

ent uptake may well be fast in small algae, but the high nutrient requirements may actually mean, that small algae need higher external nutrient concentrations to sat- isfy requirements for growth than slow-growing macroalgae. This is observed when ammonium uptake, estimated for different external ammonium concentrations and different algal species, is related to algal nitrogen requirements at maximum growth (Figure 9.7). The low nitrogen requirements of slow-growing, perennial macroalgae can be satisfied at external ammonia concentrations of about 0.5 µM, while the fast-growing *Ulva lactuca* requires an external concentration of 2 µM to achieve maximum growth. Though microalgae and ephemeral macroalgae have higher bio- mass-specific nutrient uptake, larger plants are indeed more capable of supporting maximum growth through nutrient uptake under low external concentrations.

Plant nutrient requirements can, alternatively, be at least temporarily met by ex- ploiting internal stores of nutrients [Chapman and Craigie, 1977] and it has been reported that fast-growing algae are able to build up larger nutrient stores than

slow-growing perennial algae [e.g., Lobban et al., 1985]. Though Pedersen [1993])
found that maximum nitrogen content tended to be directly related to the maximum growth rate (Figure 9.6), the critical nitrogen content exhibited a similar relationship and, therefore, the capacity to accumulate nitrogen reserves was not very
different among the six plant species of widely different form and size examined
(Figure 9.8). Because nitrogen requirements per unit time varied 50-fold, storage
capacity, expressed as the time nutrient reserves can sustain maximum growth,
was much longer for slow-growing than for fast-growing plants. Slow-growing
plants are, therefore, able to buffer long-term (e.g. seasonal) fluctuations in external nitrogen availability, while nitrogen stores in planktonic microalgae can only
sustain maximum growth for about one day without external nitrogen supplies
(Figure 9.8).

Some perennial plants, such as laminarians and seagrasses, grow by continuously re-
newing their photosynthetic tissues. This growth form allows an efficient recycling of
nutrients within the plant. Remobilization of nutrients in old tissues and subsequent reallocation to young tissues further reduce the demand for nutrient uptake
from the environment [Patriquin, 1972; Borum et al., 1989]. For a shallow-water
Danish eelgrass population, Pedersen and Borum [1993] estimated that internal recycling could account for 27% of annual nitrogen requirements and for 50%, or
more, of the requirements at the time of maximum eelgrass growth in May–June.

Consequently, slow-growing marine macrophytes seem well adapted to coastal
areas of low inorganic nutrient availability. Despite their low SA:V ratios, they are
better capable of meeting nutrient requirements because of adequate uptake mechanisms, exploitation of internal stores, and, for some plant species, internal nutrient
recycling. Hence, large macrophytes experience shorter periods of nutrient-limited
growth than small, fast-growing plants. Small plants, such as planktonic algae, depend
more on the immediate nutrient concentrations in the water column and require a
constant and rich nutrient supply to sustain a high biomass and production.

9.7. Regulation of Plant Biomass and Primary Production

Phytoplankton communities are regulated by light, nutrients and by grazing and
other loss processes (cf. chapter 4). According to the analysis above, phytoplankton
biomass and growth should be closely coupled to the total nutrient concentration in
the water column. Empirical and experimental documentation from fresh-water
lakes shows that phytoplankton biomass increases with increasing phosphorus
loading [Vollenweider, 1968; Schindler, 1978]. In marine areas, empirical relationships have been established between phytoplankton biomass and total nitrogen
concentration [Nixon and Pilson, 1983; Monbet, 1992] and phytoplankton biomass has also been experimentally stimulated by nitrogen enrichment of mesocosms [Twilley et al., 1985; Nixon et al., 1986]. In Danish coastal waters, phytoplankton biomass correlates well with concentrations of total nitrogen (corrected
for nitrogen bound in phytoplankton to eliminate autocorrelation) in the water column throughout the growth season, and in particular during summer when nitro-

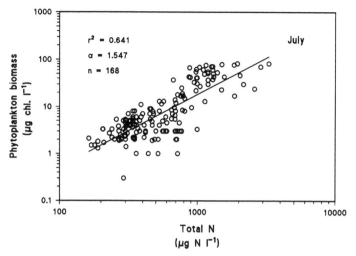

Figure 9.9. Phytoplankton biomass versus concentrations of total nitrogen (not including nitrogen bound in phytoplankton, which was subtracted by assuming a C:chlorophyll weight ratio of 40 and a molar C:N ratio of 6.6 [Redfield et al., 1963]) during summer (July). Data from different Danish coastal areas [Sand-Jensen et al., 1994a].

gen is most depleted [Sand-Jensen et al., 1994a] (Figure 9.9). Only during early spring does phytoplankton biomass correlate better with total phosphorus than with total nitrogen. Concurrent with the stimulation of phytoplankton biomass by eutrophication the total amount of suspended particles increases and both factors contribute significantly to increase light attenuation in the water column and reduce light penetration to the benthic plant communities [Sand-Jensen et al., 1994a].

The relationships between nitrogen loading or nitrogen concentration and phytoplankton biomass are, however, subject to large variability among systems [Nixon and Pilson, 1983; Monbet, 1992]. This variability may originate from differences in morphometry and physical energy input. In estuaries with a large tidal amplitude, phytoplankton biomass is significantly lower than in low-tidal estuaries when systems of equal nitrogen loading or concentrations are compared [Monbet, 1992]. Tidal influence on the amount of suspended material and, thereby, on light climate in the water column may be one explanation for this difference [Cloern, 1987; Nienhuis, 1992; Monbet, 1992], but wind-driven or tidal breakdown of stratification is another important feature [Cloern, 1984]. Phytoplankton biomass development, for example, reacts promptly to short-term changes in stratification of the shallow estuary, Roskilde Fjord [Sand-Jensen et al., 1994b]. Phytoplankton abundance increases during periods of stratification but declines during mixing, probably because the suspension-feeding mussels come into contact with the entire phytoplankton population in the mixed water column [Sand-Jensen et al., 1994b]. During stratification, the benthos can only feed on phytoplankton populations in the enclosed bottom waters. Benthic mussels or ascidians may have a sufficiently high filtration capacity to control phytoplankton biomass in shallow coastal areas

[Officer et al., 1982; chapter 10, this vol.], but water column mixing and stratification events determine when, and if, this capacity can be expressed.

Nutrient enrichment of shallow embayments stimulates growth of ephemeral macroalgae with simple morphology and high growth rates (e.g. *Ulva* [Sfriso et al., 1992], *Cladophora* [Birch et al., 1981], *Chaetomorpha* [Nienhuis, 1983]), resulting in mass accumulations of floating macroalgae in enclosed, nutrient-enriched waters [Valiela et al., 1992; Geertz-Hansen et al., 1993]. A rather complex series of interacting environmental factors such as light, temperature, nutrient concentration and herbivore grazing control the success of individual macroalgal species and, hence, species dominance and succession [Geertz-Hansen et al., 1993; Fong et al., 1993; Borum et al., 1994]. In addition, macroalgal biomass accumulation is greatly influenced by estuarine morphometry and hydrodynamics, and our ability to predict the dynamics of ephemeral macroalgal communities as a function of nutrient loading in shallow areas remains limited. However, the general pattern is that the biomass of ephemeral macroalgae increases in shallow areas, but decreases in deeper waters due to increased light attenuation from suspended material and phytoplankton [Sand-Jensen and Borum, 1991; McComb and Humphries, 1992].

Since slow-growing, perennial macrophytes seem capable of meeting their nutrient requirements even at low nutrient concentrations in the water column, nutrient enrichment may not significantly stimulate macrophyte growth. Light is a more important regulating factor, and light attenuation seems to control the depth penetration and areal cover of perennial macroalgae and seagrasses [Borum, 1983; Kautsky et al., 1986; Duarte, 1991]. Because total nutrient concentration, phytoplankton biomass, and light attenuation in the water column correlate, the lower depth limit of perennial macrophytes is an indirect function of the total nutrient concentration (Figure 9.10). In coastal Danish areas with concentrations of total nitrogen exceeding 1 mg N l^{-1} during the growth season, eelgrass, *Zostera marina*, is confined to the upper 5 m of the littoral zone while in areas with concentrations below 0.5 mg N l^{-1} eelgrass may penetrate to 10 m depth (Figure 9.10). The latter depth range corresponds to the prevailing conditions found in the outer part of Danish estuaries and open waters [Ostenfeld, 1908] before the onset of major anthropogenic nutrient input to the sea. In upper estuarine areas, such as the central basins of Limfjorden, the depth limit of eelgrass was ca. 5 m [Ostenfeld, 1908] and the median depth limit is now reduced to 2.5 m [Olesen, 1993]. According to Figure 9.10, this change in depth limit suggests that the mean concentration of total nitrogen between spring and fall has increased from 0.4 mg N l^{-1} at the beginning of this century to 0.8 mg N l^{-1} in the 1980s.

Deep growth and extensive areal cover of eelgrass and perennial macroalgae are important features for coastal ecosystems because benthic plant communities stabilize the sediment, contribute substantially to ecosystem production, and provide surfaces and shelter to a rich fauna [Mann, 1972; Thayer et al., 1975; Sand-Jensen and Borum, 1991]. Assuming that the reduction in depth penetration of macrophytes is fully reversible, should nutrient loading be reduced, the established empirical relationship between total nutrient concentration and eelgrass depth limits rep-

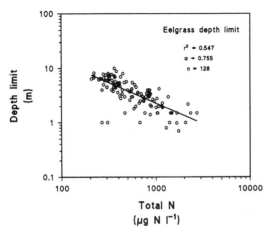

Figure 9.10. Relationship between mean concentration of total nitrogen and depth limit of eelgrass (*Zostera marina*) colonization. Data from different Danish coastal areas [Sand-Jensen et al., 1994a].

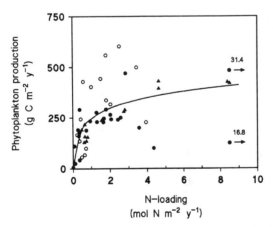

Figure 9.11. Annual phytoplankton production versus nitrogen loading of different coastal areas. Data from (O) Boynton et al. [1982], (▲) the MERL mesocosms [Nixon et al., 1986], and (●) from areas listed in Figure 9.12 and from various Danish counties. The curve represents the function: $y = 244 + 175 \log(x)$ fitted by least squares regression, r = 0.599, n = 51.

resents a powerful predictive tool for administrators managing coastal eutrophication problems. However, the relationship does not tell us anything about the time pattern of the re-establishment of former rich eelgrass beds. Such predictions rely on detailed knowledge of colonization capacity of the plants. Based on eelgrass seed production, patch formation and mortality, and subsequent vegetative patch growth, Olesen and Sand-Jensen [1994] estimated the time course of eelgrass recolonization in Limfjorden to be several decades following reduction of nutrient loading. This estimate may, however, be conservative because much faster colonization

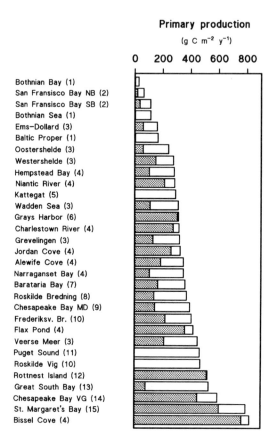

Figure 9.12. Total primary production in 31 different coastal marine areas (references in parentheses). Shaded bars: benthic production; open bars: pelagic production. References: (1) Elmgren [1984], (2) Jassby et al. [1993], (3) Nienhuis [1992], (4) Welsh et al. [1982], (5) Richardson and Christoffersen [1991], Granéli & Sundbäck [1986], (6) Thom [1984], (7) Day et al. [1973], (8) Borum et al. [1990], (9) Kemp et al. [1983], (10) Winter et al. [1975], (11) Jensen et al. [1990], (12) Walker et al. [1988], (13) Lively et al. [1983], (14) Murray & Wetzel [1987], (15) Mann [1972].

of deep-water eelgrass populations was observed in Lake Grevelingen, where plant areal cover was more than doubled within a 10-year period after an improvement of light penetration [Verhagen and Nienhuis, 1983].

Just like phytoplankton biomass increases with increasing nutrient loading so does phytoplankton production [Boynton et al., 1982]. Estimates of annual pelagic production from different coastal areas and from the MERL mesocosm experiments [Nixon et al., 1986] show that nitrogen loading stimulates phytoplankton production (Figure 9.11). However, the increase in production levels off at high loading rates and maximum rates of production rarely exceed 500 g C m^{-2} yr^{-1}. This pattern, firstly, reflects a gradual decrease in community turnover (growth rate) with

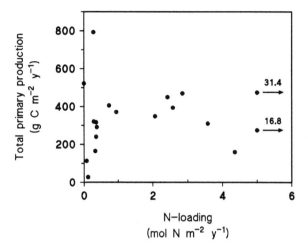

Figure 9.13. Total annual primary production versus nitrogen loading of different coastal areas. Data from those areas listed in Figure 9.12 where the N-loading is known [Borum and Sand-Jensen, in press].

increasing nutrient concentration, and, secondly, a decline in the efficiency by which nutrients are utilized by phytoplankton communities. Apparently, the controlling role of nutrients for annual phytoplankton production gradually diminishes with total nutrient concentration and is replaced by other constraining factors such as light climate and loss processes.

Though phytoplankton production of coastal areas is stimulated by nutrient enrichment, total primary production (including contributions from benthic plant communities) may not be enhanced, because the increase in phytoplankton biomass reduces depth penetration and areal cover of benthic plant populations, i.e. the ecosystem is light-limited. Perennial kelps and seagrass meadows may attain annual production rates of 1,000 g C m^{-2} or more [Mánn, 1972; McRoy and McMillan, 1977; Pedersen and Borum, 1993]. Commonly, the integrated organic carbon production of entire estuarine or coastal ecosystems ranges between 250 and 500 g C m^{-2} yr^{-1} (Figure 9.12) [Nixon et al., 1986] and, apparently, total nutrient concentration has little impact on systems production [Borum and Sand-Jensen, in press] (Figure 9.13). Pristine coastal areas may exhibit high rates of annual production originating primarily from dense macrophyte beds and with low contributions by phytoplankton. In regions subject to high anthropogenic loading, phytoplankton production may reach the 500 g C m^{-2} yr^{-1} but contributions from benthic plants are then negligible. Hence, the most important effect of nutrient enrichment on primary producers in shallow coastal areas is to induce qualitative changes in plant community composition and primary production [Borum and Sand-Jensen, in press]. Integrated primary production does not seem to be affected to a great extent, though an enhancement of primary production may occur in deeper coastal areas that are phytoplankton dominated under unperturbed conditions (cf. chapter 5).

9.8. Influence on Oxygen Dynamics and Higher Trophic Levels

If total organic carbon production of coastal systems remains rather unaffected by nutrient enrichment, what is then the influence of eutrophication on oxygen dynamics and higher trophic levels within the coastal zone? There seem to be no obvious reasons for expecting major changes in net system production with increased nutrient loading of coastal areas. Nixon and Pilson [1984] reported that Narragansett Bay exhibited net export of organic matter, but coastal areas are most often slightly heterotrophic at the systems level [e.g., Smith et al., 1991] indicating that autochtonously produced organic matter is primarily mineralized within the coastal zone itself irrespective of nutrient loading. Therefore, annual oxygen consumption and release should be independent of anthropogenic loading, and we can hypothesize that observed changes in oxygen dynamics must originate from increased spatial and temporal separation of oxygen release and consumption rather than increased rates. The changes in plant composition described above could, theoretically, induce increased separation. However, the complex relationships between plant dominance and oxygen dynamics in shallow coastal waters need to be examined more thoroughly.

Likewise, the influence of coastal zone nutrient enrichment on faunal composition and secondary production seems to be rather complex (see chapter 8). There are examples, especially from deeper phytoplankton dominated areas, suggesting that the biomass of benthic animals increases with increasing nutrient loading (cf. chapter 8), but, in general, shallow coastal waters cannot be shown to respond to eutrophication by producing a larger biomass of animals [Nixon et al., 1986]. Nixon et al. [1986] suggested that the lack of observed responses can be partly explained by inadequate harvesting and measuring techniques. Alternatively, however, the lack of response could reflect that total organic carbon production may not be enhanced. Hence, eutrophication effects on higher trophic levels may predominantly influence species composition of animal populations (cf. chapter 10) rather than total secondary production.

Acknowledgments. A large number of people have contributed or helped in various ways to obtain the data and form the discussion of this chapter. I am in particular grateful to the other members of our research group at the Freshwater-Biological Laboratory, K. Sand-Jensen, O. Geertz-Hansen, M. F. Pedersen, O. Pedersen, M. Hein, and B. Kjøller, for providing theories, data, discussions, constructive criticism and technical support. I also thank M. Johnson, and representatives from the counties of Aarhus, Nordjylland, Viborg and Ringkøbing for letting me utilize unpublished data in this review.

References

Ahlgren, I., Limnological studies of Lake Norrviken, a eutrophicated Swedish lake, I: Water chemistry and nutrient budget, *Hydrobiologia, 29,* 53–90, 1967.

Aller, R. C., Benthic fauna and biogeochemical processes in marine sediments, in *Nitrogen Cycling in coastal Marine Environments,* edited by T. H. Blackburn and J. Sørensen, pp. 301–338, John Wiley & Sons, Chichester, 1988.

Andersen, J. M., Nitrogen and phosphorus budgets and the role of sediments in six shallow Danish lakes, *Arch. Hydrobiol., 74*, 528–550, 1971.

Atkinson, M. J., and S. V. Smith, C:N:P ratios of benthic marine plants, *Limnol. Oceanogr., 28*, 568–574, 1984.

Billen, G., C. Lancelot, and M. Meybeck, N, P, and Si retention along the aquatic continuum from land to ocean, in *Ocean Margin Processes in Global Change*, edited by R. F. C. Mantoura, J.-M. Martin and R. Wollast, pp. 19–44, John Wiley & Sons, Chichester, 1991.

Birch, P. B., D. M. Gordon, and A. J. McComb, Nitrogen and phosphorus nutrition of *Cladophora* in the Peel-Harvey system, *Bot. Mar., 24*, 381–387, 1981.

Borum, J., The quantitative role of macrophytes, epiphytes, and phytoplankton under different nutrient conditions in Roskilde Fjord, Denmark, *Proc. Intl Symp. Aquat. Macrophytes*, pp. 35–40, Nijmegen, 1983.

Borum, J., Development of epiphytic communities on eelgrass (*Zostera marina*) along a nutrient gradient in a Danish estuary, *Mar. Biol., 87*, 211–218, 1985.

Borum, J., and K. Sand-Jensen, Is total primary production in shallow coastal marine waters stimulated by nitrogen loading?, Oikos, in press.

Borum, J., L. Murray, and W.M. Kemp, Aspects of nitrogen acquisition and conservation in eelgrass plants, *Aquat. Bot., 35*, 289–300, 1989.

Borum, J., O. Geertz-Hansen, K. Sand-Jensen, and S. Wium-Andersen, Eutrofiering – effekter på marine primærproducenter, *NPO-forskning fra Miljøstyrelsen*, C3, 52 pp., Copenhagen, 1990.

Borum, J., M. F. Pedersen, L. Kær, and P. M. Pedersen, Growth and nutrient dynamics in marine plants (in Danish), *Havforskning fra Miljøstyrelsen, 41, 59* pp., Danish Environmental Protection Agency, Copenhagen, 1994.

Boynton, W. R., W. M. Kemp, and C. W. Keefe, A comparative analysis of nutrients and other factors influencing estuarine phytoplankton production, in *Estuarine Comparisons*, edited by V. S. Kennedy, pp. 69–90, Academic Press, New York, 1982.

Caffrey, J. M., and W. M. Kemp, Nitrogen cycling in sediments with estuarine populations of *Potamogeton perfoliatus* and *Zostera marina*, *Mar. Ecol. Prog. Ser., 66*, 147–160, 1990.

Cambridge, M. L., A. W. Chiffings, C. Brittan, L. Moore, and A. J. McComb, The loss of seagrass in Cockburn Sound, Western Australia. II. Possible causes of seagrass decline, *Aquat. Bot., 24*, 269–285, 1986.

Chapman, A. R. O., and J. S. Craigie, Seasonal growth in *Laminaria longicruris*: relations with dissolved inorganic nutrients and internal reserves of nitrogen, *Mar. Biol., 40*, 197–205, 1977.

Cloern, J. E., Temporal dynamics and ecological significance of salinity stratification in an estuary (South San Francisco Bay, USA), *Oceanol. Acta, 7*, 137–141, 1984.

Cloern, J. E., Turbidity as a control on phytoplankton biomass and productivity in estuaries, *Cont. Shelf Res., 7*, 1367–1381, 1987.

Day, J. W. Jr., W. G. Smith, P. R. Wagner, and W. C. Stowe, Community structure and carbon budget of a salt marsh and shallow bay estuarine system in Louisiana, *Publ. No. LSU-SG-72-04*, 79 pp., Louisiana State University, Baton Rouge, 1973.

Duarte, C. M., Seagrass nutrient content, *Mar. Ecol. Prog. Ser., 67*, 201–207, 1990.

Duarte, C. M., Seagrass depth limits, *Aquat. Bot., 40*, 363–372, 1991.

Duarte, C. M., Nutrient concentration of aquatic plants: patterns across species, *Limnol. Oceanogr., 37*, 882–889, 1992.

Duarte, C. M., Submerged aquatic vegetation in relation to different nutrient regimes, *Intl Symp. Nut. Dyn. Coast. Estuar. Envir.*, pp. 87–112, Helsingør, Denmark, 1995.

Elmgren, R., Trophic dynamics in the enclosed, brackish Baltic Sea, *Rapp. P.-v. Réun. Intl Explor. Mer., 183*, 152–169, 1984.

Floderus, S., Sedimentation and resuspension in Aarhus Bay (in Danish), *Havforskning fra Miljøstyrelsen, 18*, 39 pp., Danish Environmental Protection Agency, Copenhagen, 1992.

Fong, P., R. M. Donohoe, and J. B. Zedler, Competition with macroalgae and benthic cyanobacterial mats limits phytoplankton abundance in experimental microcosms, *Mar. Ecol. Prog. Ser.*, 100, 97–102, 1993.

Fyns Amt, *Det Fynske Vandmiljø 1992*, 40 pp., County of Fyn, 1993.

Geertz-Hansen, O., K. Sand-Jensen, D. F. Hansen, and A. Christiansen, Growth and grazing control of abundance of the marine macroalga, *Ulva lactuca* L. in a eutrophic Danish estuary, *Aquat. Bot.*, 46, 101–109, 1993.

Granéli, W., and K. Sundbäck, Can microbenthic photosynthesis influence below-halocline oxygen conditions in the Kattegat?, *Ophelia, 26*, 195–206, 1986.

Granéli, E., W. Granéli, and L. Rydberg, Nutrient limitation at the ecosystem and the phytoplankton community level in the Laholm Bay, south-east Kattegat, *Ophelia, 26*, 181–194, 1986.

Hanisak, M. D., Nitrogen limitation of *Codium fragile* spp. *tomentosoides* as determined by tissue analysis, *Mar. Biol., 50*, 333–337, 1979.

Harlin, M. M., and B. Thorne-Miller, Nutrient enrichment of seagrass beds in a Rhode Island coastal lagoon, *Mar. Biol., 65*, 221–229, 1981.

Hattori, A., Denitrification and dissimilatory nitrate reduction, in *Nitrogen in the Marine Environment*, edited by E. J. Carpenter and D. G. Capone, Academic Press, New York, 1983.

Hein, M., M. F. Pedersen, and K. Sand-Jensen, Size-dependent nitrogen uptake in micro- and macroalgae, *Mar. Ecol. Prog. Ser.*, 118, 247–253, 1995.

Jassby, A. D., J. E. Cloern, and T. M. Powell, Organic carbon sources and sinks in San Francisco Bay: variability induced by river flow, *Mar. Ecol. Prog. Ser.*, 95, 39–54, 1993.

Jenkins, M. C., and W. M. Kemp, The coupling of nitrification and denitrification in two estuarine sediments, *Limnol. Oceanogr.*, 29, 609–619, 1984.

Jensen, L. M., K. Sand-Jensen, S. Marcher, and M. Hansen, Plankton community respiration along a nutrient gradient in a shallow Danish estuary, *Mar. Ecol. Prog. Ser.*, 61, 75–85, 1990.

Kamp-Nielsen, L., Benthic-pelagic coupling of nutrient metabolism along and estuarine eutrophication gradient, *Hydrobiologia, 235/236*, 457–470, 1992.

Kautsky, N., H. Kautsky, U. Kautsky, and M. Waern, Decreased depth penetration of *Fucus vesiculosus* (L.) since the 1940's indicates eutrophication of the Baltic Sea, *Mar. Ecol. Prog. Ser.*, 28, 1–8, 1986.

Kemp, W. M., R. R. Twilley, J. C. Stevenson, and L. G. Ward, The decline of submerged vascular plants in upper Chesapeake Bay: Summary of results concerning possible causes, *Math. Tech. Soc. J.*, 17, 78–89, 1983.

Kristensen, E., Benthic fauna and biogeochemical processes in marine sediments: microbial activities and fluxes, in *Nitrogen Cycling in Coastal Marine Environments*, edited by T. H. Blackburn and J. Sørensen, pp. 275–299, John Wiley & Sons, Chichester, 1988.

Lively, J. S., Z. Kaufman, and E. J. Carpenter, Phytoplankton ecology of a barrier island estuary: Great South Bay, New York, *Estuar. Coast. Shelf Sci.*, 16, 51–68, 1983.

Lobban, C. S., P. J. Harrison, and M. J. Duncan, *The Physiological Ecology of Seaweeds*, 242 pp., Cambridge University Press, Cambridge, 1985.

Mann, K. H., Ecological energetics of the sea-weed zone in a marine bay on the Atlantic coast of Canada II. Productivity of the seaweeds, *Mar. Biol., 14*, 199–209, 1972.

McComb, A. J., and R. Humphries, Loss of nutrients from catchments and their ecological impacts in the Peel-Harvey estuarine system, Western Australia, *Estuaries, 15*, 529–537, 1992.

McRoy, C. P., and C. McMillan, Production ecology and physiology of seagrass, in *Handbook of Seagrass Biology*, edited by C.P. McRoy and C. Helfferich, pp. 53–87, Marcel Dekker, New York, 1977.

Monbet, Y., Control of phytoplankton biomass in estuaries: A comparative analysis of microtidal and macrotidal estuaries, *Estuaries, 15*, 563–571, 1992.

Murray, L., and R. L. Wetzel, Oxygen production and consumption associated with the major autotrophic components in two temperate seagrass communities, *Mar. Ecol. Prog. Ser., 38*, 231–239, 1987.

Nielsen, L. P., P. B. Christensen, and S. Rysgaard, Denitrification in coastal waters and fjords (in Danish), *Havforskning fra Miljøstyrelsen, 50*, 49 pp., Danish Environmental Protection Agency, Copenhagen, 1994.

Nienhuis, P. H., Temporal and spatial patterns of eelgrass (*Zostera marina* L.) in a former estuary in The Netherlands, dominated by human activities. *Mar. Tech. Soc. J., 17*, 69– 77, 1983.

Nienhuis, P. H. Eutrophication, water management, and the functioning of Dutch estuaries and coastal lagoons, *Estuaries, 15*, 538–548, 1992.

Nixon, S. W., and M. E. Q. Pilson, Nitrogen in estuarine and coastal marine ecosystems, in *Nitrogen in the Marine Environment*, edited by E. J. Carpenter and D. G. Capone, pp. 565–648, Academic Press, New York, 1983.

Nixon, S. W., and M. E. Q. Pilson, Estuarine total system metabolism and organic exchange calculated from nutrient ratios: An example from Narragansett Bay, in *The Estuary as a Filter*, edited by V. S. Kennedy, pp. 261–290, Academic Press, New York, 1984.

Nixon, S. W., C. A. Oviatt, J. Frithsen, and B. Sullivan, Nutrients and the productivity of estuarine and coastal marine ecosystems, *J. Limnol. Soc. Sth Afr., 12*, 43–71, 1986.

Odum, E. P., *Fundamentals of Ecology*, 574 pp., W. B. Saunders, Philadelphia, 1971.

Officer, C. B., T. J. Smayda,, and R. Mann, Benthic filter feeding: A natural eutrophication control, *Mar. Ecol. Prog. Ser., 9*, 203–210, 1982.

Olesen, B., Bestandsdynamik hos ålegræs, Ph.D. thesis, 94 pp., Department of Plant Ecology, University of Aarhus, 1993.

Olesen, B., and K. Sand-Jensen, Patch dynamics of eelgrass, *Zostera marina, Mar. Ecol. Prog. Ser., 166*, 147–156, 1994.

Ostenfeld, C. H., On the ecology and distribution of the grass-wrack (*Zostera marina*) in Danish waters, in *Report of Danish Biological Station to the Board of Agriculture, XVI*, pp. 1–62 Copenhagen, 1908.

Patriquin, D. G., The origin of nitrogen and phosphorus for growth of the marine angiosperm *Thalassia testudinum, Mar. Biol., 15*, 35–46, 1972.

Pedersen, M. F., Vækst og næringsstofdynamik hos marine planter, Ph.D thesis, 99 pp., Freshwater-Biological Laboratory, University of Copenhagen, 1993.

Pejrup, M., J. Bartholdy, and A. Jensen, Supply and exchange of water and nutrients in the Grådyb tidal area, Denmark, *Estuar. Coast. Shelf. Sci., 36*, 221–234, 1993.

Pelegrí, S. P., L. P. Nielsen, and T. H. Blackburn, Denitrification in estuarine sediment stimulated by irrigation activity of the amphipod *Corophium volutator, Mar. Ecol. Prog. Ser., 105*, 285–290, 1994.

Redfield, A. C., B. A. Ketchum, and F. A. Richards, The influence of organisms on the composition of sea-water, in *The Sea, Vol. II*, edited by M. N. Hill, pp. 26–77, Wiley, London, 1963.

Revsbech, N. P., B. B. Jørgensen, and O. Brix, Primary production of microalgae in sediments measured by oxygen microprofile, $H^{14}CO_3^-$ fixation and oxygen exchange methods, *Limnol. Oceanogr., 26*, 717–730, 1981.

Richardson, K., and A. Christoffersen, Seasonal distribution and production of phytoplankton in the southern Kattegat, *Mar. Ecol. Prog. Ser., 78*, 217–227, 1991.

Ringkøbing Amtskommune, *Rapport om Næringssalte og Vandskifte i Ringkøbing Fjord*, 12 pp., Torben Larsen Hydraulics Aps, 1992.

Risgaard-Petersen, N., S. Rysgaard, L. P. Nielsen,, and N. P. Revsbech, Diurnal variation of denitrification in sediments colonized by benthic microphytes. *Limnol. Oceanogr., 39*, 573–579, 1994.

Ryther, J. H., and W. H. Dunstan, Nitrogen, phosphorus, and eutrophication in the coastal, marine environment, *Science, 171*, 1008–1013, 1971.

Sand-Jensen, K., and J. Borum, Interactions among phytoplankton, periphyton, and macrophytes in temperate freshwaters and estuaries, *Aquat. Bot., 41*, 137–176, 1991.

Sand-Jensen, K., S. L. Nielsen, J. Borum, and O. Geertz-Hansen, Phytoplankton and macrophyte development in Danish coastal waters (in Danish), *Havforskning fra Miljøstyrelsen, 30*, 43 pp., Danish Environmental Protection Agency, Copenhagen, 1994a.

Sand-Jensen, K., J. Borum, O. Geertz-Hansen, J. N. Jensen, A. B. Josefson, and B. Riemann, Resuspension and biological interactions in Roskilde Fjord (in Danish), *Havforskning fra Miljøstyrelsen, 51*, 69 pp., Danish Environmental Protection Agency, Copenhagen, 1994b.

Schindler, D. W., Factors regulating phytoplankton production and standing crop in the world's fresh waters, *Limnol. Oceanogr., 23*, 478–486, 1978.

Seitzinger, S. P., S. W. Nixon, and M. E. Q. Pilson, Denitrification and nitrous oxide production in a coastal marine ecosystem, *Limnol. Oceanogr., 29*, 73–83, 1984.

Seitzinger, S. P., Denitrification in freshwater and coastal marine ecosystems: Ecological and geochemical significance, *Limnol. Oceanogr., 33*(4), 702–724, 1988.

Seitzinger, S. P., Denitrification in aquatic sediments, in *Denitrification in Soil and Sediment*, edited by N. P. Revsbech and J. Sørensen, pp. 301–322, Plenum Press, New York, 1990.

Sfriso, A., B. Pavoni, A. Marcomini, and A. A. Orio, Macroalgae, nutrient cycles, and pollutants in the Lagoon of Venice, *Estuaries, 15*, 517–528, 1992.

Sharp, J. H., J. R. Pennock, T. M. Church, J. M. Tramontano, and L. A. Cifuentes, The estuarine interaction of nutrients, organics, and metals: A case study in the Delaware Estuary, in *The Estuary as a Filter*, edited by V. S. Kennedy, pp. 241–258, Academic Press, New York, 1984.

Smith, R. E. H., and J. Kalff, Size-dependent phosphorus uptake kinetics and cell quota in phytoplankton, *J. Phycol., 18*, 275–284, 1982.

Smith, S. V., J. T. Hollibaugh, S. J. Dollar, and S. Vink, Tomales Bay metabolism: C:N:P stoichiometry and ecosystem heterotrophy at the land –sea interface, *Estuar. Coast. Shelf Sci., 33*, 223–257, 1991.

Thayer, G. W., D. A. Wolfe, and R. B. Williams, The impact of man on seagrass systems, *Am. Sci., 63*, 288–296, 1975.

Thom, R. M., Primary production in Grays Harbor Estuary, Washington, *Bull. S. Calif. Acad. Sci., 83*, 99–105, 1984.

Twilley, R. R., W. M. Kemp, K. W. Staver, J. C. Stevenson, and W. R., Boynton, Nutrient enrichment of estuarine submersed vascular plant communities: I. Algal growth and effects on production of plants and associated communities, *Mar. Ecol. Prog. Ser., 23*, 179–191, 1985.

Valiela, I., K. Foreman, M. LaMontagne, D. Hersh, J. Costa, P. Peckol, B. DeMeo-Andreson, C. D'Avanzo, M. Babione, C.-H. Sham, J. Brawley, and K. Lajtha, Couplings of watersheds and coastal waters: Sources and consequences of nutrient enrichment in Waquoit Bay, Massachusetts, *Estuaries, 15*, 443–457, 1992.

Verhagen, J. H. G., and P. H. Nienhuis, A simulation model of production, seasonal changes in biomass and distribution of eelgrass (*Zostera marina*) in Lake Grevelingen, *Mar. Ecol. Prog. Ser., 10*, 187–195, 1983.

Vollenweider, R. A., *Scientific Fundamentals of the Eutrophication of Lakes and Flowing Waters, with Particular Reference to Nitrogen and Phosphorus as Factors in Eutrophication*, 192 pp., OECD, DAS/CSI/68.27, Paris, 1968.

Walker, D. I., R. J. Masini, and E. I. Paling, Comparison of annual production and nutrient status of the primary producers in a shallow limestone reef system (Rottnest Island), Western Australia, Australian Marine Science Association Silver Jubilee 1963–1988 Annual Conference, 13–16 December, pp. 183–187, University of Sydney, Sydney, 1988.

Welsh, B. L., R. B. Whitlatch, and W. F. Bohlen, Relationship between physical characteristics and organic carbon sources as a basis for comparing estuaries in southern New England, in *Estuarine Comparisons*, edited by V. S. Kennedy, pp. 53–67, Academic Press, New York, 1982.

Westlake, D. F., Comparisons of plant productivity, *Biol. Rev.*, 38, 385–425, 1963.

Winter, D. F., K. Banse, and G. C. Andersen, The dynamics of phytoplankton blooms in Puget Sound, a fjord in the Northwestern United States, *Mar. Biol.*, 29, 139–176, 1975.

Wulff, F., A. Stigebrandt, and L. Rahm, Nutrient dynamics of the Baltic Sea, *Ambio, 19*, 126–133, 1990.

10

Case Study: Kertinge Nor

Hans Ulrik Riisgård, Carsten Jürgensen and Frede Østergaard Andersen

10.1. Introduction

This case study describes investigations conducted in 1991–1992 in Kertinge Nor, a shallow cove on the northern part of Fyn, Denmark (Figure 10.1). By the end of 1989, annual discharges of nitrogen (N) and phosphorus (P) to the fjord system Kertinge Nor/Kerteminde Fjord were reduced by 45% and 78%, respectively, as domestic sewage was no longer led into the system.

The significant reduction in the external nutrient load made the fjord system suitable for studying how nutrient reduction affected its recovery from eutrophication. Kertinge Nor combines exceptionally clear water, dominance of macrophytes (eelgrass, *Zostera marina*, and thick mats of filamentous algae, *Chaetomorpha linum*), high densities of small jellyfish (*Aurelia aurita*) and a dense population of benthic ascidians (*Ciona intestinalis*). This made the cove an interesting area to study the dynamics of the biological structure in an eutrophic ecosystem in which nutrient fluxes and suspension-feeding organisms play a decisive role.

This chapter presents the main results of concurrent projects on hydrography, nutrients, macrophytes, plankton, jellyfish, and ascidians and synthesizes the interactions between nutrient dynamics and the biological structure.

10.2. Water Exchange

The fjord system consisting of Kerteminde Fjord and Kertinge Nor (Figure 10.1) covers an area of 8.5 km² and has a mean water depth of 2 m and a maximum depth of 8 m. The fjord has a sill at its mouth to the open sea (Great Belt). The discharge over the sill is forced by a diurnal tide with an average amplitude of ca. 20 cm. The tide gives rise to maximum discharges at the fjord entrance of 100–200 m³ s⁻¹.

Eutrophication in Coastal Marine Ecosystems
Coastal and Estuarine Studies, Volume 52, Pages 205–220
Copyright 1996 by the American Geophysical Union

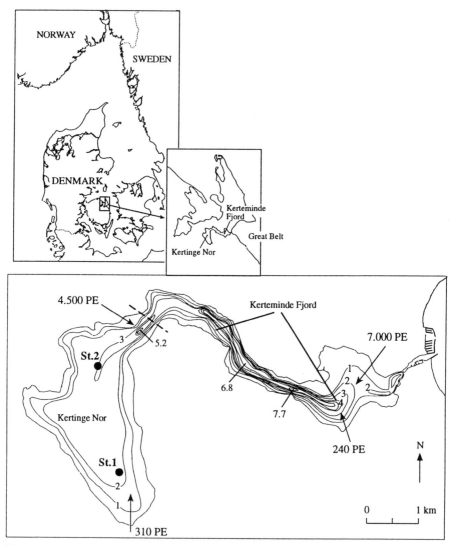

Figure 10.1. Investigation area, Kertinge Nor/Kerteminde Fjord, Denmark. Water depths (m) are shown. Sampling stations and amounts of discharged sewage (PE = person equivalents) to the fjord-system in 1989 (before sewage reduction) are indicated.

The fresh-water input of 0–0.05 m^3 s^{-1} is negligible with respect to the water exchange of the fjord system. The salinity in the central part of the system varies typically between 14 and 22 psu over the year. The temperature ranges between 0 and 22°C [Larsen et al., 1994].

The water exchange of the fjord system is governed by density-driven circulation. The salinity in the Great Belt outside the fjord varies as a result of changing flow

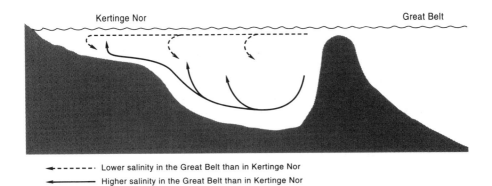

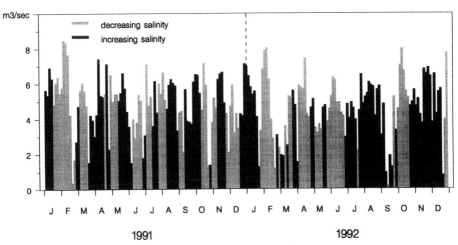

Figure 10.2. Upper part of figure: Illustration of water exchange in Kertinge Nor (KN). Low salinity in the Great Belt (GB) compared to KN gives rise to a density-driven circulation that flows into the fjord system at the surface and out at the bottom (counter-clockwise). High salinity in GB compared to KN gives rise to a density-driven circulation that flows into the fjord system at the bottom and out at the surface (clockwise). Lower part of figure: Calculated flow rates of the density-driven exchange of water between the Kertinge Nor and the Great Belt. The light bars indicate flow situations with decreasing salinity in the Great Belt, i.e. counter-clockwise circulation, and the dark bars indicate increasing salinity, i.e. clockwise circulation. From Jürgensen [1995].

situations (cf. chapter 3). Outflow of water from the Baltic Sea gives salinities down to 10 psu, whereas inflow to the Baltic Sea gives salinities up to 27 psu in the upper layer of the Great Belt. Because saline water is more dense than fresh water, the salinity variations cause longitudinal density variations from the inner part of the fjord system to the mouth. As a consequence of longitudinal density gradient, density-driven vertical circulation occurs. When tidal forcing flushes dense water over the sill it will flow down below the fjord water and give rise to a density-driven circulation system within the entire fjord system. When, on the other hand, lighter water is forced into the fjord the circulation is in the opposite direction. On an annual time scale the two circulation directions have equal probability. An approximate residence time of water in the central areas of the system is between one week and a few months, with an average of approximately 1.5 month [Jürgensen, 1995]. However, because of the dynamics of the exchange processes, the "residence time" is not a very useful concept in this fjord system. A qualitative illustration of the flow rate and the direction of the exchange of water is given in Figure 10.2.

10.3. Nitrogen and Phosphorus Balance

A general transport pattern of nutrients through the fjord system was obtained from studies on total nitrogen (tot-N) and total phosphorus (tot-P). The transport of the nutrients was described by means of the numerical, hydrodynamic model system "MIKE11". The model solves the equations for conservation of mass and momentum and describes the advective and the dispersive transport of matter [DHI, 1990]. The model was calibrated and verified by salinity [Jürgensen, 1995]. The transport of total phosphorus through the fjord entrance was measured directly and was in close agreement with the simulations.

The daily nutrient inputs from point sources, diffuse sources and the atmosphere were calculated based on direct measurements. The nutrient fluxes from the sediment to the pelagic were in the model determined as the difference between the computed and the measured nutrient concentrations. The nutrient fluxes from the sediment estimated by the model were confirmed by laboratory incubations (Christensen et al., 1994]. The incubations showed the same temporal variations and the same order of magnitude as the numerical simulations.

After the sewage discharge to the fjord system had been stopped by the end of 1989, the annual nutrient load was reduced by 45% N and 78% P. Consequently, the release of nutrients from the sediment (internal load) became much more important for the total nutrient balance of the fjord system. During the summer periods in 1991 and 1992 the sediment was the dominating source for both tot-N and tot-P. The nutrient fluxes calculated from the model, however, also showed periods with low nutrient uptake by the sediment (Figure 10.3).

In 1991, the annual release of P from the sediment was 3.3 times higher than the total external input from land and air, while the annual net release of N was only 0.4 times the external load (Table 10.1). In 1992, when the nutrient-assimilating mat of filamentous algae disappeared (see later), the sediment release was 3.4 times and 1.9 times higher for P and N, respectively, than the external input. The extremely high release of N and P observed during late June 1992 (Figure 10.3) coincided with the decay of the thick filamentous algal mat.

The model also showed a net export of both N and P from the fjord to the open sea (Figure 10.3). Apart from short periods with minor imports in the spring and in the autumn, tot-N was exported over the entire two-year period. For tot-P, a minor

TABLE 10.1. Balance of nitrogen (N) and phosphorus (P) in Kertinge Nor/Kerteminde Fjord in 1991 and 1992. All figures are in metric tonnes. From Christensen et al. [1994].

| | 1991 | | 1992 | |
	N	P	N	P
Input from land and precipitation	35	1.6	40	1.6
Net release from sediment	13	5.4	75	5.5
Net export to the Great Belt	48	7	115	7.1

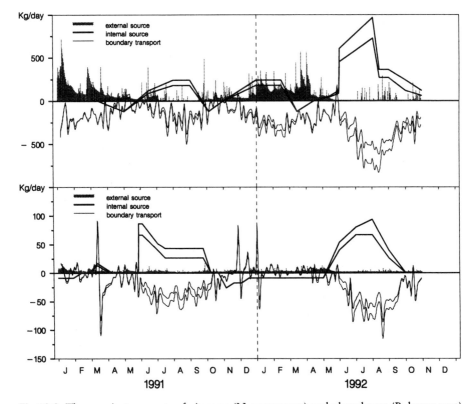

Fig 10.3. Three main transports of nitrogen (N, upper part) and phosphorus (P, lower part) in Kertinge Nor.

External source: Nutrient loads from point sources (waste water), diffuse sources (agriculture), and atmospheric deposition (rain).

Internal source: Nutrient exchanges between water and sediment, positive values for release from sediment and negative values for uptake.

Boundary transport: Nutrient transports over the boundary between Kertinge Nor and the Great Belt. Positive values are imports into the Kertinge Nor, negative values are exports to the Great Belt.

The two lines for internal sources and boundary transports give the upper and lower estimates of the simulations. From Jürgensen [1995].

import was observed in the winter, whereas there was a pronounced export during the rest of the period. The net export of P from the fjord system to the Great Belt was approximately 7 tonnes in both 1991 and 1992, whereas the export of N was 48 tonnes in 1991 (a "normal year" with extensive growth of benthic plants) and 115 tonnes in 1992. The loss of N through denitrification in the fjord system was estimated to 26 tonnes N in 1992. The total loss of P and N from the system may be compared to the mobile pools of these nutrients in the sediment, which were estimated to be 30–52 tonnes of P and 150–225 tonnes of N.

10.4. Nutrient Concentrations

The concentration of tot-N varied throughout the season from 25 to 150 μmol l^{-1}, with characteristic peaks in concentration during winter and late summer (Figure 10.4a). The winter peaks reflect fresh-water runoff from the surrounding area, which is the major external input of inorganic nitrogen (inorg-N) to the system. The late summer peaks were due to sediment release of inorg-N which was rapidly assimilated by the phytoplankton. The fast assimilation resulted in very low inorg-N concentrations during the summer months (Figure 10.4a). The tot-N concentration was significantly higher during 1992 compared to the preceding years.

The concentration of tot-P varied between 0.5 and 12 μmol l^{-1}, with the lowest values in December–March and the highest in May–October (Figure 10.4b). The same pattern was observed for inorg-P with concentrations of up to 6 μmol l^{-1} during the summer due to internal loading.

The spatial and temporal variation in inorg-P and tot-P in 1991 was followed at 13 stations in the fjord system and compared with a station in the Great Belt [Møhlenberg and Jürgensen, 1994]. Longitudinal gradients of inorg-P with the lowest concentrations in the inner part of the fjord system indicated that Kertinge Nor acted as a sink for inorg-P during late February to mid April, probably due to assimilation by the macrophytes. From late May through October the opposite gradient was found, indicating that Kertinge Nor acted as a source for inorg-P and tot-P. In November and December the concentration of inorg-P again decreased from the Great Belt toward Kertinge Nor implicating an inward flux of inorg-P, probably driven by adsorption and precipitation processes in the sediment. The measured gradients are in close agreement with concentrations computed from the tot-P release simulations.

10.5. Macrophytes

In the southern part of Kertinge Nor (St. 1 on Figure 10.1), a dense mat of filamentous green algae (primarily *Chaetomorpha linum*) covered the sediment surface with maximum biomasses of 65 g d.wt m^{-2} in August 1991 [Larsen et al., 1994]. In 1992, the biomass of these filamentous algae peaked already in early June (80 g

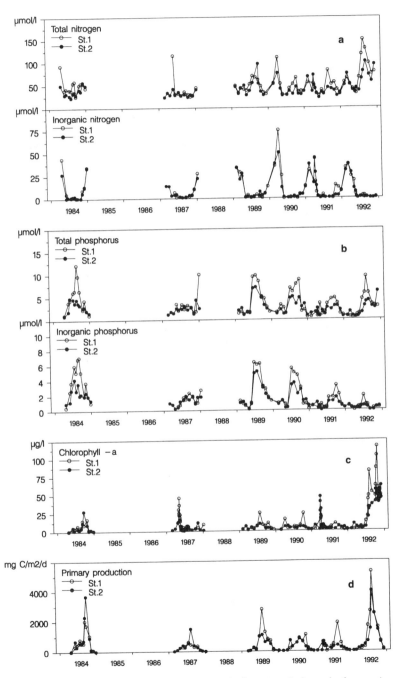

Figure 10.4. Nitrogen (a), phosphorus (b), chlorophyll-*a* (c) and phytoplankton primary production (d) in Kertinge Nor in 1984, 1987 and 1989–1992. From Riisgård et al. [1995].

d.wt m^{-2}), after which time they started dying. No plants were found in August and throughout the rest of 1992.

Profile measurements through the mat in May 1992 (while the algae were still growing well) showed high O_2 concentrations in the upper half of the mat but anoxic conditions in the lower part [Christensen et al., 1994]. Due to the anoxic conditions high NH_4^+ and PO_4^{3-} concentrations were found at the sediment surface underneath the mat. The concentration of these nutrients decreased up through the mat because of assimilation by the algae. The steep nutrient gradient thus indicated that the mat was very efficient as a filter for the nutrients released from the sediment. However, the nutrients bound in the algal biomass were released rather quickly as the algae died and decayed in 1992.

The bottom flora of the northern part of Kertinge Nor (St. 2 on Figure 10.1) was dominated by rooted macrophytes, mainly eelgrass, but filamentous algae were also found between the shoots of eelgrass. It was expected that the eelgrass beds would increase and that the coverage of filamentous algae could decrease as a consequence of reduced nutrient loading and lower concentrations in the water because eelgrass is able to satisfy its nutrient demand from the sediment through root uptake.

A study on the influence of eelgrass and filamentous algae on the cycling of P revealed that the P release from sediments under filamentous algae was 2.5 times higher than from the sediment covered with eelgrass [Christensen et al., 1994]. The release of P from *Chaetomorpha* detritus was also significantly higher than from eelgrass detritus (decay constants: 0.090 and 0.029 d^{-1}, respectively). These results indicate that increases in the eelgrass beds at the expense of filamentous algae will have a positive effect on the recovery of the fjord system because nutrients will be less available for the phytoplankton during the growth season.

10.6. Plankton Dynamics

The biomass of phytoplankton, expressed as the water column concentration of chlorophyll-*a*, was normally low during most years (Figure 10.4c). Intensive sampling during February and March suggested that there was a short spring maximum of chlorophyll-*a* (Figure 10.4c). Apart from this peak in early spring, chlorophyll-*a* never exceeded 10 µg l^{-1} during 1991, which was similar to the preceding years (Figure 10.4c). The chlorophyll-*a* concentration was also low during the first five months of 1992, but in July chlorophyll-*a* increased markedly with concentrations reaching 120 µg l^{-1} in October. The phytoplankton primary production also increased markedly during the summer of 1992 and the rates were significantly higher than in the preceding "normal" years (Figure 10.4D).

The pelagic biomass levels and the successions of plankton species were similar during the spring periods of 1991 and 1992, but the developments were very different during the two summer and autumn periods (Figure 10.5). In 1991 the auto- and heterotrophic biomasses were low and dominated by diatoms and dinophyceans as well as ciliates, rotifers and epibenthic harpacticoids (Table 10.2). In 1992 the bio-

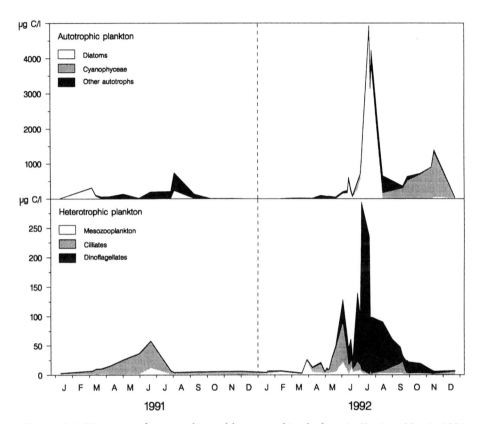

Figure 10.5. Biomasses of autotrophic and heterotrophic plankton in Kertinge Nor in 1991 and 1992. Based on data from Riisgård et al. [1995].

TABLE 10.2. Mean heterotrophic biomasses (μg C l^{-1}) on stations 1 and 2 in Kertinge Nor (see Figure 10.1) during the summer period (May–September) 1991 and 1992. From Riisgård et al. [1995].

| | 1991 | | 1992 | |
	St. 1	St. 2	St. 1	St. 2
Dinoflagellates	<0.1	10.3	54.6	87.5
Ciliates	6.3	1.4	12.4	11.8
Copepods*	2.5	0.4	1.8	2.7
Rotifers	<0.1	<0.1	1.0	4.2
Spionid larvae	0.6	0.1	1.1	0.7
Bivalve larvae	<0.1	0.1	0.2	0.1
Total	9.4	12.3	81	107

* mainly harpacticoids

mass of phytoplankton was exceptionally high. In the beginning of the period the autotrophic biomass was dominated by the diatom, *Skeletonema costatum*, followed later by another diatom, *Stephanodiscus hantzschii*. The diatom bloom was succeeded by a bloom of small cyanobacteria which lasted until the end of November. The heterotrophic biomass was dominated by ciliates in the beginning of the period, but was later succeeded by heterotrophic dinoflagellates appearing in very high biomasses.

10.7. Jellyfish (*Aurelia aurita*)

The predation impact by the jellyfish, *Aurelia aurita*, on zooplankton in Kertinge Nor was studied during 1991 and 1992. The water processing capacity (clearance) as a function of medusa size and water temperature was measured in the laboratory using rotifers (*Brachionus plicatilis*) or copepods (*Acartia tonsa*) as prey organisms. From these data and from measurements of medusa density and size distribution in the Kertinge Nor the clearance capacities of the jellyfish population were estimated [Olesen et al., 1994; Olesen, 1995].

In 1992 the first young medusae (ephyrae) of the new generation of *Aurelia aurita* appeared in February. During the following months, numbers increased dramatically and a maximum density of about 300 individuals m^{-3} was measured in April. From April and throughout the rest of the season, the density of medusae decreased (Figure 10.6).

The growth pattern of *Aurelia aurita* was different in 1991 compared with 1992. During 1991 the growth was poor until early August, when a pronounced increase in both umbrella diameter and specific growth rate was observed (Figure 10.6). A maximum in the umbrella diameter was observed in early September 1991, where a mean of 54 mm was measured. The instantaneous specific growth rate (μ) reached a maximum of 0.07 d^{-1} in late August 1991. In 1992 a maximum mean umbrella diameter of only 37 mm was measured in late June, where also a maximum of μ = 0.09 d^{-1} was found.

Small medusae (umbrella diameter = 4 mm) collected in Kertinge Nor showed rapid growth in the laboratory when rotifers were offered in the concentration range of 130–13,000 rotifers l^{-1}. During a 10-day incubation period, the umbrella diameter increased from approximately 4 to 9 mm at prey concentrations higher than 700 rotifers l^{-1}. This corresponded to a dry-weight increase from 0.1 to 0.9 mg. At the lowest concentration of rotifers (130 ind. l^{-1}) the diameter increased to 6.5 mm, corresponding to an increase of 0.36 mg. A maximum instantaneous specific growth rate of 0.22 d^{-1} was obtained at a prey density of approximately 400 rotifers l^{-1}, corresponding to a growth of 60 μg C l^{-1}.

The maximal specific growth rate of 0.22 d^{-1} found in the laboratory experiments may be compared to the maximal growth rate of 0.09 d^{-1} observed in Kertinge Nor where the mean carbon concentration of the zooplankton was only 6.5 and 4.7 μg C l^{-1} in 1991 and 1992, respectively. It can thus be concluded that *Aurelia*

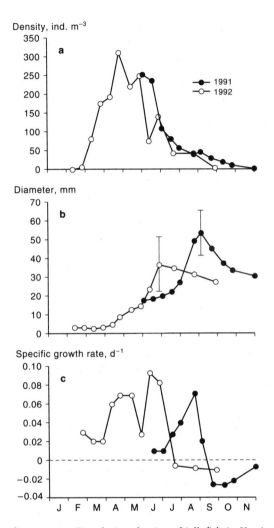

Figure 10.6. *Aurelia aurita*. a: Population density of jellyfish in Kertinge Nor in 1991 and 1992. b: Mean diameter of jellyfish; representative s.d. is shown for two dates. c: Instantaneous specific growth rate of jellyfish. [From Olesen et al., 1994.]

aurita in Kertinge Nor was severely food-limited and never exploited its potential for growth.

The estimated population clearance capacities showed that *Aurelia aurita* was able to filter a water volume corresponding to several times the whole water volume of Kertinge Nor daily during August and September 1991, as well as in June 1992. Furthermore, the mean residence time $(t_{1/2})$ for a zooplankton organism was < 5 days from spring to autumn. This predation impact explains the low biomass of zooplankton during part of the zooplankton growth season.

TABLE 10.3. *Ciona intestinalis*. Population density (95% confidence limits are given in parentheses), individual total dry weight and population filtration rate at different times of the year in Kertinge Nor. The theoretical mean residence time ($t_{1/2}$) of an algal cell is shown as well as the quotient (Q) of total population filtration potential of ascidians covering 2.81 km² to the whole water volume in Kertinge Nor (5.48 km² with an average depth of 2 m). Based on Petersen and Riisgård [1992]; revised according to Riisgård et al. [1995].

Time	Temperature, °C	Density, ind. m⁻²	Weight, mg ind.⁻¹	Population filtration rate		$t_{1/2}$	Q
				l h⁻¹ m⁻²	×10⁴ m³ h⁻¹	h	d⁻¹
March 1991	6.4	76 (45–129)	57	30	8.4	46	0.2
May 1991	13.5	20 (13–31)	169	38	11	37	0.2
Sep.1991	12.4	237 (195–287)	56	191	54	7	1.2
Dec. 1991	3.5	277 (207–372)	49	47	13	30	0.3
April 1992	8.6	139 (113–170)	90	104	29	13	0.6
June 1992	20.2	33 (25–44)	295	139	39	10	0.9

10.8. Ascidians (*Ciona intestinalis*)

In order to establish the potential grazing impact of *Ciona intestinalis* in Kertinge Nor the population densities and the individual filtration rates as a function of size of the ascidians, water temperature and algal cell concentration were determined [Petersen and Riisgård, 1992].

The maximum filtration rate of *Ciona intestinalis* (F_{max}, ml min⁻¹) expressed as a function of the total body dry weight (W, g) at a given temperature (T, °C) was described as $F_{max} = (8.3T - 6.9)W^{0.68}$. Thus, from population estimates and water temperature in Kertinge Nor, population filtration could be computed (Table 10.3). It is seen that population filtration varied considerably during the year. In September 1991 the population could clear a volume of water corresponding to the total volume of Kertinge Nor per day, while in May 1991 it could only clear 0.2 times the total volume per day. To illustrate the potential grazing impact of the total *C. intestinalis* population on phytoplankton it was calculated that the theoretical mean residence time ($t_{1/2}$) for algal cells in a fully mixed water column (2 m³) varied between 7 and 46 h, depending on time of year (Table 10.3).

The filtration capacity of the *Ciona* population in Kertinge Nor varied both during the year and between years as a result of variation in population density and individual temperature- and size-dependent filtration rates. In general, the filtration capacity was low in winter, due to low temperatures, and in mid-summer, due to low population biomass, while it was high in late summer and early fall, where temperatures and biomass were both high.

No data on *Ciona* population density were obtained from July to December 1992 due to high water turbidity, but the filtration rate was decreased 8–10 fold due to overloading of the feeding system by high in situ concentrations of the diatom,

Stephanodiscus hantzschii, which dominated the phytoplankton in July 1992. Thus, reduced filtration rates diminished the grazing impact by *C. intestinalis* in the summer and fall of 1992.

10.9. Biological Structure and Nutrient Dynamics

The biological structure of the Kertinge Nor ecosystem in 1991 is summarized in Figure 10.7. The water column was extremely clear which allowed sufficient light penetration to the bottom where a significant benthic primary production of filamentous algae and eelgrass took place. The dense algal mat was important for the control of the nutrient flux from the sediment into the water column. Below the algal mat the sediment was black and sulfidic due to anoxic conditions and without living animals. On the algal mat, however, a large number (3000–4000 ind. m^{-2}) of small (< 3 mm) cockles, *Cardium* sp., were observed clinging to the algal filaments just above the anoxic bottom. Furthermore, small snails, *Littorina* sp. (4000–5000 ind. m^{-2}) and *Hydrobia* sp. (2000–3000 ind. m^{-2}), were seen together with mussels, *Mytilus edulis* (60–80 ind. m^{-2}; < 20 mm shell length). Thus, under the prevailing conditions all the benthic animals had moved up above the anoxic zone.

During summer 1991, the water processing capacity of the jellyfish population was very high, with a maximum rate obtained in early September where the jellyfish population could daily process a water volume corresponding to ca. 13 times the whole water volume of Kertinge Nor. This suggests that *Aurelia aurita* could well control zooplankton in Kertinge Nor during summer and fall. Laboratory experiments proved that the medusae were food-limited at in situ zooplankton concentrations. Moreover, *A. aurita* was apparently growing in excess of its food source, as zooplankton densities in the water column of Kertinge Nor during the day (Table 10.2) could not explain the observed growth of *A. aurita*. It has more recently been observed, however, that the density of harpacticoids in the water column during night can exceed the density during the day by a factor of 20 (unpublished results) and night-swimming harpacticoids may therefore have been an important food source for the jellyfish in Kertinge Nor in 1991.

In 1991 the filter-feeding *Ciona intestinalis* exerted a high grazing pressure on phytoplankton, which partly explained the low observed phytoplankton biomass. In particular, during late summer and fall, the *Ciona* population reached densities of ca. 250 ind. m^{-2}. During fall, the dense population of *C. intestinalis* had the potential capacity to filter the total water volume of Kertinge Nor 0.2–1.2 times daily, and the mean residence time of an algal cell in the water column ($t_{1/2}$) was only 7 h in September 1991 (Table 10.3). From previous data and observations this situation seemed to have been the prevailing for the structure and dynamics of Kertinge Nor for the decade preceding 1992.

During 1991 and the previous "normal" years the water column of Kertinge Nor was very clear and the chlorophyll-*a* concentrations were low throughout the growth season (Figure 10.4c). At the same time, a high biomass of the filamentous

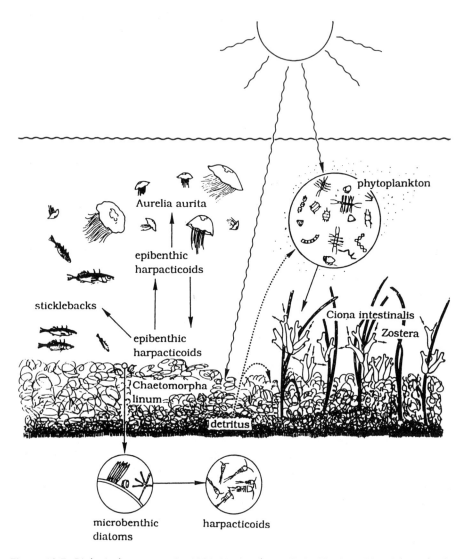

Figure 10.7. Biological structure in 1991 ("normal year") in Kertinge Nor. Three food-chains were identified: 1) phytoplankton → ascidians (*Ciona intestinalis*); 2) epiphytic diatoms → epibenthic harpacticoids → jellyfish (*Aurelia aurita*) + sticklebacks; 3) macrophytes → detritus → decomposing microorganisms. From Riisgård et al. [1995].

macroalga, *Chaetomorpha linum*, was recorded on the bottom throughout the year (mats up to 40 cm thick during the summer). The production of filamentous algae was limited by the availability of N, and the algal mat therefore controlled the release of inorganic N from the sediment [Christensen et al., 1994]. This "normal" situation existed for Kertinge Nor until June 1992. A long period of calm clear weather with high insolation began in the middle of May and persisted for 8–9 weeks. This caused high primary production in the filamentous algal mat which led to the formation of large oxygen bubbles that caused parts of the whole algal mat to float up from the sediment to the water surface [Christensen et al., 1994]. Floating in the water, the filamentous algal mat could no longer act as an absorbing filter for the nutrient flux from the underlying anoxic sediment and large amounts of nutrients were released to the water column.

The phytoplankton population immediately responded to the elevated nutrient concentrations. Within a week, the phytoplankton biomass increased 80-fold in areas where the algal mat was more or less separated from the sediment surface [Christensen et al., 1994]. During June 1992, the phytoplankton biomass increased almost exponentially which reduced the light penetration depth into the water to only 30 cm. The massive shading by phytoplankton caused a total collapse of the remaining benthic algal mats and from August *Chaetomorpha linum* had disappeared from Kertinge Nor. The growth of eelgrass also decreased. This enabled a high flux of nutrients from the sediment (Figure 10.3) and extremely high chlorophyll-*a* concentrations were measured in October (Figure 10.4c).

During June 1992 the parent generation of *Ciona intestinalis* died off due to its natural life cycle. The filtration capacity of the ascidian population was, thus, low by the end of June and beginning of July because only newly settled specimen were present. The decline in filtration capacity by the end of June 1992 was coincident with the increase in phytoplankton biomass. In July algal cell concentrations exceeded optimum concentrations for filtration and *C. intestinalis* became saturated and subsequently reduced its filtration rate to a minimum of its potential capacity [Petersen and Riisgård, 1992]. *C. intestinalis* therefore did not control phytoplankton during the summer and fall of 1992 in Kertinge Nor.

The marked change in primary producers had no significant influence on the jellyfish population which occurred in comparably high densities during both 1991 and 1992. Also, the water processing rates were approximately equal during both years and the high predation rate prevented the mesozooplankton from increasing in numbers as a response to the increased phytoplankton biomass (Table 10.2). Thus, even during extreme situations with chlorophyll-*a* concentrations of up to 120 μg l^{-1}, the *Aurelia aurita* population had an important regulating impact on the zooplankton density.

It may be concluded that Kertinge Nor until 1992 was an inherently unstable eutrophic ecosystem in which interactions between suspension-feeding organisms and mobilization of nutrients from the sediment determined the dynamics of the biological structure. In the present case study the instability of the system blurred all possible short-term effects caused by the significant reduction in nutrient discharge by the end of 1989.

The estimated export rates of nutrients (Table 10.1) may be expected to lead to a marked reduction of the internal nutrient load within the next 5–10 years, though the exhaustion rates for nutrients from the sediment pools (enhanced by former sewage disposal) will decrease exponentially over the years. The exhaustion of nutrients may also be dependent on the frequency of "abnormal years". Possible alteration of the biological structure in Kertinge Nor during this reestablishing period remains unknown at the present, and no reliable information exists about the former, unpolluted fjord system. The first environmental examination of the system was made in 1974 by the county of Fyn, but at that time the system was already eutrophic.

Following the present case study, the county of Fyn has been responsible for monitoring the development in Kertinge Nor. So far, it can be stated that the phytoplankton biomass was low during both 1993 and 1994 ("normal years" with clear water; see Figure 10.4). The filamentous algal mat was very weakly developed, whereas the eelgrass had favorable growth conditions. Also the *Ciona intestinalis* population was dense. Furthermore, a normal brackish-water infauna had developed extensively. At the present time it is unknown whether the Kertinge Nor ecosystem, possibly due to the radical interruptions of the "normal" biological structure and massive nutrient loss in 1992, has started a development toward a new, more stable condition.

References

Christensen, P. B., F. Møhlenberg, D. Krause-Jensen, H. S. Jensen, S. Rysgaard, P. Clausen, O. Sortkjær, L. Schlüter, S. B. Josefsen, C. Jürgensen, F. Ø. Andersen, J. Thomassen, M. S. Thomsen, and L. P. Nielsen, Nutrient transport and biological interactions in Kertinge Nor/Kerteminde Fjord (in Danish), 128 pp., *Havforskning fra Miljøstyrelsen, 43*, Danish Environmental Protection Agency, Copenhagen, 1994.

DHI, MIKE11, A microcomputer based modelling system for rivers and channels, *Technical Reference*, Danish Hydraulic Institute, Hørsholm, Denmark, 1990.

Jürgensen, C. Modelling of nutrient release from the sediment in a tidal inlet, Kertinge Nor, Funen, Denmark, *Ophelia, 42*, 163–178, 1995.

Larsen, G. R., C. Jürgensen, P. B. Christensen, N. J. Olesen, J. K. Petersen, J. N. Jensen, E. Mortensen, O. Sortkjær, P. Andersen, Kertinge Nor/Kerteminde Fjord – status and development (in Danish), *Havforskning fra Miljøstyrelsen, 44*, 132 pp., Danish Environmental Protection Agency, Copenhagen, 1994.

Møhlenberg, F., and C. Jürgensen, Spatial and temporal variation in phosphorus in a small marine inlet after cut-off of sewer discharges, in *Changes in Fluxes in Estuaries: Implications from Science to Management*, edited by K. Dyer, and B. Orth, pp. 215–218, Olsen & Olsen, Fredensborg, Denmark. 1994.

Olesen, N. J., K. Frandsen, and H. U. Riisgård, Population dynamics, growth and energetics of jellyfish (*Aurelia aurita*) in a shallow fjord, *Mar. Ecol. Prog. Ser., 105*, 9–18, 1994.

Olesen, N. J., Clearance potential of the jellyfish *Aurelia aurita*, and predation impact on zooplankton in a shallow cove, *Mar. Ecol. Prog. Ser., 124*, 63–72, 1995.

Petersen, J. K., and H. U. Riisgård, Filtration capacity of the ascidian *Ciona intestinalis* and its grazing impact in a shallow fjord, *Mar. Ecol. Prog. Ser., 88*, 9–17, 1992. (see also Erratum: *Mar. Ecol. Prog. Ser., 108*, 204).

Riisgård, H. U., I. Clausen, F. Møhlenberg, J. K. Petersen, N. J. Olesen, P. B. Christensen, M. M. Møller, and P. Andersen, Filter-feeding animals, plankton dynamics and biological structure in Kertinge Nor (in Danish), *Havforskning fra Miljøstyrelsen, 45*, 75 pp., Danish Environmental Protection Agency, Copenhagen, 1994.

Riisgård, H. U., P. B. Christensen, N. J. Olesen, J. K. Petersen, M. M. Møller, and P. Andersen, Biological structure in a shallow cove (Kertinge Nor, Denmark) – Control by benthic nutrient fluxes and suspension-feeding ascidians and jellyfish, *Ophelia, 41*, 329–344, 1995.

11

The Use of Models in Eutrophication Studies

André W. Visser and Lars Kamp-Nielsen

11.1. Introduction

Models are different things to different people. They are essential to the manner in which we as humans view the world; a concept recognized by Greek philosophers 3000 years ago. They are also fundamental to scientific understanding. The essential character of a model is to simplify reality in some way with the express purpose of highlighting or accentuating certain relationships and processes. Models can be conceptual and expressed verbally (dialogues and discourses of renaissance science). They can be physical or a diagram; scaled up or down to bring them to a human dimension (colored balls connected by straws to represent molecules, points of light projected onto the ceiling of a darkened auditorium to represent the celestial sphere). In modern scientific usage, the term model has generally come to refer to a mathematical description.

Mathematical models have for a long time been an integral part of scientific practice in the hard sciences such as physics and chemistry and in their application in engineering. Their unquestionable success lies primarily in the limited number of universal laws and constants which appear to govern most of the interactions of matter and energy. In contrast, ecosystems are extremely complex and the principles behind their behavior remain largely unknown. Nevertheless, the descriptive and predictive powers of mathematical models have attracted biologists and ecological researchers over the past decades.

The range of these models has covered a broad spectrum of reality [Platt et al., 1981] from *reductionist*: examining specific processes isolated from external influences, a 'mathematical lab experiment' if you will to *holistic*: attempting to cover as many interconnected processes as is feasible in simulating a natural system. Models

Eutrophication in Coastal Marine Ecosystems
Coastal and Estuarine Studies, Volume 52, Pages 221–242
Copyright 1996 by the American Geophysical Union

can also be categorized by the type of question they seek to address [Franks, 1995]. They can be *theoretical* in that they qualitatively explore the interaction of well-founded formulations in an innovative manner leading to new testable hypotheses. They can be *heuristic*, formulated and parameterized with regard to a well-defined set of observations to gain insight into the dynamics of a particular system. Finally, they can be *predictive*, extrapolating a heuristic model beyond its sample range. In practice, models have characteristics of all three categories. They have a theoretical basis which, together with observations, is used to parameterize a heuristic model, the predictive power of which is used to gauge the success of the underlying concepts and formulation.

Lastly, models can be typified by their scope. These include *empirical* models, sometimes called black box models, which seek to describe an observed data set as a mathematical function usually through a statistical fitting procedure disregarding any possible causal processes. Such models are useful for identifying trends, as test beds against which conceptual models can be formulated (i.e. causal processes might be inferred from the mathematical function derived), and can be used as predictive models for 'well-behaved' systems. In the case of fresh-water systems, an empirical approach [Vollenweider, 1968, 1976] has provided the successful basis for predicting and managing eutrophication. Unfortunately very few marine systems are 'well-behaved' and empirical models can fail dramatically when applied by the unwary. *Dynamic* models on the other hand seek to build up the behavior of a system from well-defined subprocesses and their supposed interrelation. If empirical models try to describe *how* things change, dynamic models investigate *why* they change. While dynamic models are difficult to implement, they do have a clear advantage that in principle they can respond to changing input and forcing conditions. Their predictive capability is more soundly based on natural processes than that of empirical models. In oceanography, the unqualified term 'model' is most often used to refer to a dynamic rather than an empirical description.

In whichever manner models are classified, they are artifices, designed by people for a specific purpose. The inherent conflict between mathematical tractability and realism leads to compromises between simplicity and complexity depending on the purpose for which the model is used. Scientists should use the model in the hypothetico-deductive process characterizing the 'strong interference' scientific method [Platt, 1964]. This is needed in ecosystem research since observed phenomena can often be explained by different theories and be caused by different mechanisms [Scheffer and Beets, 1994], a consequence of the usual non-linearity of these systems. In this context, models can be used to get an overview of complex structures and for testing and/or falsifying functional relationships. Consequently, the model is equally valuable if it acts unexpectedly and a hypothesis can be rejected as when it behaves as expected. For such strategic purposes, models should be as simple as possible to be penetrable by analytic techniques; model results should not be as confounding as direct observation.

Once models have been verified and validated they can serve as predictive tools for decision makers and managers, provided that the predictions made are within the range of validation. The model, for these tactical purposes, is most credible when it acts as expected.

It is generally accepted that study of eutrophication in limnic systems is well in advance of that in marine systems. This is partly because eutrophication in lakes was recognized earlier, but also because it was simpler to quantify. Lakes are morphologically well-defined systems with measurable in- and outflows. More importantly, the flushing rates of lakes are generally much slower than their mixing rates so that it can be readily assumed that similar conditions apply throughout. While much can be learned by marine managers and scientists from the expertise accrued by limnologists over the years, it should be stressed that a direct transfer is inappropriate; the marine environment presents a new set of challenges. Marine systems have much faster flushing rates, often similar to their mixing rates as well as the time scale over which pertinent biological processes operate. While flushing rates can be made 'artificially' slower by considering a larger area, this makes the system less well defined. It is a characteristic of marine systems that fluctuations dominate and mean conditions are seldom encountered. Interpretation of observations is complicated: does a change in some parameter indicate a change in the system, or is it a transient pulse from outside. In marine systems, the detection of a trend requires years of data, the detection of a change in trend requires decades.

Given the poor quantification of eutrophication (as defined by Nixon [1995]; a change in trophic state) and its effects in marine systems, few eutrophication modeling studies as such have been carried out. However, there is a growing body of work investigating ecosystem models in marine environments. These models contain, as an integral part, descriptions of coupled nutrient and biological processes. Further, it is becoming apparent that physical processes are particularly important in determining key characteristics of various marine systems. This realization has led to the development of dynamic physical-biological-chemical ecosystem models [Legendre and Demers, 1984]. Since it is likely that the potential of such models will become increasingly exploited in the near future in investigating eutrophication, we will devote some time in what follows to discuss their formulation.

The purpose of this chapter is to present some key concepts and methods for eutrophication studies in marine systems which are useful or might prove so in the near future. In this we choose a few examples from lakes, estuaries, fjords and coastal seas as illustrations. We do not intend this as an exhaustive review of the subject.

11.2. Model Construction

Once a problem or hypothesis has been identified through management/policy needs or scientific interests, the construction of a mathematical model can proceed. Generally this is along well-proven lines [Nihoul, 1976] as illustrated Figure 11.1. The first step is a *demarcation* of the system into physical space; where and when and state variables; what. For instance, physical space may represent a lake, estuary, fjord or coastal sea over a seasonal cycle. In eutrophication studies, at least

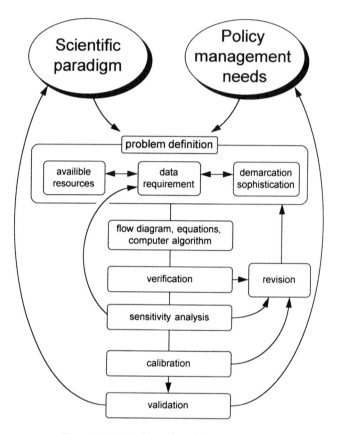

Figure 11.1. Outline of model construction.

some of the state variables would be nutrients (N, P, and others). However, there are other variables which may be modeled and potentially play an important role in nutrient concentrations in a particular system, e.g. phytoplankton, zooplankton, detritus etc. Demarcation of the system is in recognition that a model cannot be all-encompassing and encapsulates the compromise that must be struck between simplicity and reality.

The second step is given this demarcation, how are *external influences* on the system treated. These include boundary conditions, initial conditions and forcing functions. In mathematical terms this ensures that the model is well posed. Boundary conditions describe how state variables are influenced from outside the model domain, for instance the flux of nutrients supplied to an estuary by river inflow. Likewise, initial conditions determine the pre-existing state of the system and the importance of its history dependence. Those influences deemed important to the system but which cannot be dynamically modeled must be imposed as external forcing functions. An example here is the annual variation of light intensity governing photosynthesis in mid-latitude coastal seas.

Finally, mathematical equations are used to express the *dynamic interactions* of the state variables. In this context, it is often useful to sketch the interaction between dynamic state variables in a flow diagram, e.g. Figure 11.6 illustrates the flow of nitrogen between state variables in an idealized fjord model. A fundamental law exploited in eutrophication, indeed in ecological modeling studies is that of mass balance. If C is the concentration of some nutrient in a system then the time rate of change of C is given by:

$$dC/dt = F_{in} - F_{out} + R \qquad (1)$$

where F_{in} and F_{out} are the fluxes in and out of the system, and R is the net increase due to internal production and removal (if removal is greater then production then R is negative). Fluxes in and out of the system are generally governed by two physical mechanisms, diffusion (or mixing) and advection (see chapter 3). R is generally a biological term expressing the transfer to and from other state variables.

In all but the most reductionist models, these dynamic interactions are non-linear which must be solved by numerical rather than analytic means. Numerical solution techniques require a *discretization* of the governing dynamic equations. That is, solutions can only be specified at particular points in space and time. An important consequence of this is the treatment of sub-grid processes. For example a *sub-grid* process common to nearly all environmental models is turbulence, or more specifically how turbulence affects the mixing of material and momentum. Discretization also has implications to technical considerations such as numerical stability.

Once implemented as a computer simulation, a model goes trough a series of stages:

- Verification: the model does what is expected.
- Sensitivity analysis: how detailed do the governing parameters and forcing functions need to be. Since for operational and tactical applications, forcing functions have to be supplied by a real-time monitoring program, this is a critical step for managers to consider in determining the cost effectiveness of implementing a model.
- Calibration: those parameters which are poorly known from nature can be 'tuned' to give the best fit of the model to observations. While commonly practiced in the past, tuning is becoming unfashionable in the scientific community. In complex models there may be scores of parameters, any combination of which can be tinkered with to get a 'good fit'. Generalist models should be tuned as little as possible: it is better to have a discrepancy between model results and observations than between the parameters used and those measured experimentally.
- Validation: testing the model output against independent observations. Inherent in this step is defining a criterion. If the model is to be used as a management tool, those involved as end users should contribute to the establishment of this validation criterion.

Model development is an iterative process involving refinement as new resources, data, and understanding become available. Valuable models eventually lead to interpretations useful for policy and management implementation and in exceptional cases to a review of the scientific paradigm upon which they are based.

11.3. Lakes

Early this century, much effort in limnology was devoted to classification and grouping of lakes according to their productivity as defined from their species composition. However, many limnologists felt lake typology was a blind alley and thus turned to statistical analysis and experimental ecology. The eutrophication of lakes by phosphorus made limnology focus on quantifying nutrient loading, its relationship to concentrations and primary production, and modeling these relationships.

One class is the simple empirical steady-state models, built on multi-lake, cross-sectional analysis which considers lakes as continuously stirred tank reactors, much like a chemostat in chemical engineering. Behind all eutrophication models lies the law of conservation of mass which for a nutrient in a confined body of water can be written

$$d(CV)/dt = M - QC - S \qquad (2)$$

where C is the nutrient concentration, V the volume of the water body, M the incoming mass of nutrient, Q the discharge from the water body, and S the mass of nutrient lost to sedimentation or the atmosphere. If the incoming nutrient is biologically available and taken up by phytoplankton, then the amount sedimented out depends on the amount of detritus produced either directly from phytoplankton or through higher trophic levels. In this case it can be assumed that S is proportional to mass of nutrient in the lake, CV, with a proportionality factor α, termed the loss coefficient. At steady state, Eq. 2 can be rewritten as

$$C = L_c/(Hr)\,(r/(r+\alpha)) = L_c(1 - R_c)/(Hr) \qquad (3)$$

where $r = Q/V$ is the flushing rate, $L_c = HM/V$ is the area loading of nutrient, H is the mean depth of the water body, and $R_c = \alpha/(r + \alpha)$ is termed the retention coefficient. The net loss of nutrient expressed either as α or R_c and its relationship to site-specific constants like morphology and hydraulic regime has challenged limnologist for more than 25 years. In a review, Kristensen et al. [1990] tested 20 of these empirical models for inlake phosphorus concentrations P in relation to the loading on 131 annual data sets from Danish lakes. Among the best was the well-known OECD-model application [Vollenweider and Kerekes, 1982] which parameterizes the phosphorus retention coefficient as:

$$R_p = 1 - 1.55\,P\exp(-0.18)/(1 + r\exp 0.5)\exp(0.82) \qquad (4)$$

In a similar vein, Jensen et al. [1995] developed and tested an empirical model for nitrogen concentration and loading for 98 Danish lakes.

Parallel to the development of these nutrient loading-concentration models, have been models relating nutrient concentrations to phytoplankton biomass in lakes [e.g. Nicholls and Dillon, 1978; Ahlgren et al., 1988]. The principal behind these is Lieberg's law of limiting factors, that the limiting nutrient is taken up as a constant fraction of the cell mass or pigment content. While attractive as a simplifying principal, many laboratory and field experiments have demonstrated that these conditions are not always met. For instance different nutrients can be limiting for different components of the phytoplankton community at the same time, cell volume and pigment content vary with nutrient status and light climate, luxury uptake of P and N can occur at high concentrations, other autotrophic and heterotrophic components of the ecosystem can compete for nutrients, and chemical reactions are also possible. Nevertheless, for concentration ranges of total phosphorus up to 3 μM, correlations with summer chlorophyll are fair with linear slope of 0.5 – 1 [Ahlgren et al., 1988].

These successes notwithstanding, the complexity, difficulties in parameterization and the possibility of multiple causality have led to the role of empirical models being questioned. In response, limnic modelers are turning to theoretical, dynamic models with high-resolution temporal and spatial scales, to explore the relationship between environmental factors and eutrophication.

11.3. Estuaries

Estuaries are the most immediate marine systems to feel the impact of anthropogenic loading of nutrients and organic material. An estuary is by definition an area where sea water and fresh water and their associated properties mix. However, unlike rivers flowing into lakes, the fresh water entering an estuary has a dynamic influence on circulation and mixing by virtue of its lower salinity and density (see chapter 3).

The mixing of fresh and salt water in the estuary means that the inflow is not only from the river, but also from the adjacent sea (Figure 11.2). Conservation of volume and salt within an estuary give:

$$Q_R + Q_B = Q_S \quad \text{and} \quad S_R Q_R + S_B Q_B = S_S Q_S$$

which can be rearranged to give

$$Q_S = Q_R (S_B - S_R)/(S_B - S_S) \quad \text{and} \quad Q_B = Q_R (S_S - S_R)/(S_B - S_S)$$

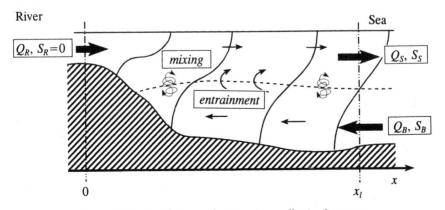

Figure 11.2. Circulation and mixing in a well-mixed estuary.

[e.g. Bowden, 1984]. For instance, the river Rhine has a mean discharge rate of Q_R = 1200 m³/s and carries essentially no salt, $S_R = 0$ psu. At the mouth of the Rhine, the entrance to the Rotterdam waterway, mean surface and bottom salinities are 27 psu and 30 psu, respectively. This gives a mean surface outflow, and bottom inflow of $Q_S = 12,000$ m³/s and $Q_B = 10,800$ m³/s. That is, an overwhelming percentage of the water found in the Rhine Estuary originates from the North Sea rather than from the river itself.

This is reflected in the flushing rate r of the estuary;

$$r = (Q_R + Q_B) / V = (S_B - S_R)/(S_B - S_S) \, Q_R / V \qquad (5)$$

which for the Rhine is 10 times faster than the flushing rate for a lake of the same volume with the same river inflow: $r' = Q_R / V$. The energy required to move this enormous amount of water comes not from the hydraulic head of the river, but rather from the potential energy inherent in the fresh–salt water partition.

One useful property exploited in estuarine studies is that salt is a conservative tracer. This concept was used by Officer [1979] to examine the behavior of non-conservative dissolved constituents (e.g. nutrients) empirically. Briefly, the one-dimensional (along estuary) steady-state advection diffusion equations for salt S and a nutrient C at some location can be written

$$Q_R S - K_x A \, dS/dx = 0 \qquad (6)$$

$$Q_R C - K_x A \, dC/dx = F \qquad (7)$$

where A is the cross-sectional area of the estuary, K_x is the longitudinal dispersion coefficient, and x is directed positive down-estuary. $x = 0$ is the river end, and $x = l$ is the sea end of the estuary. F is the flux of the nutrient at this particular location and can, through manipulation of Eqs 6 and 7, be written as:

$$F = Q_R (C - S \, dC/dS) \qquad (8)$$

The net loss of nutrient C from the estuary is then given by $L = F(S_0) - F(S_l)$ which, if the salinity at the river end of the estuary is $S_0 = 0$ gives a fractional loss G written as

$$G = (C_0 - C_0^*) / C_0 , \quad \text{where } C_0^* = C_l - S_l (dC/dS)|_l \qquad (9)$$

Thus, plotting the mean salinity versus mean nutrient concentration along an estuary gives an empirical relationship of the utilization or release of the nutrient (Figure 11.3) within the estuary without specific dependence on river discharge or mixing characteristics. While this method has found some refinement [e.g. Officer and Lynch, 1981], it remains the basis for a powerful empirical tool in eutrophication studies [e.g. Officer and Ryther, 1980] in examining the role of silicate depletion in limiting primary production [van Raaphorst et al., 1988] in examining the regeneration of benthic phosphorus in eutrophication of a shallow water estuary).

Dynamic mass balance models of estuaries and other marine systems involving the mixing of two distinctive water masses are wide spread [e.g. Grantham and Tett, 1981; Smith and Veeh, 1989; Pejrup et al., 1993; Hansen et al., 1995]. This is largely

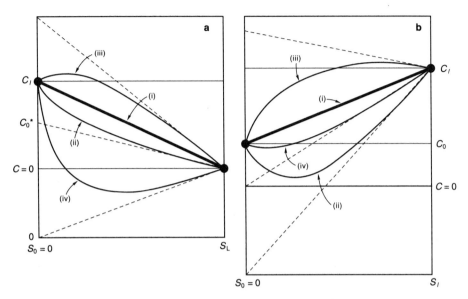

Figure 11.3. The fractional loss G of a nutrient concentration C in an estuary as deduced from C versus salinity S relationship. a: concentration in river C_0 > concentration in sea C_1; (i) conservative ($G = 0$), (ii) non-conservative and estuary acts as sink ($G < 1$), (iii) non-conservative and estuary acts as source ($G > 1$), (iv) no net flux out of estuary ($G = 1$). b: concentration in river C_0 < concentration in sea C_1; (i) conservative, (ii) non-conservative and estuary acts as sink, (iii) non-conservative and estuary acts as source, (iv) no net flux into estuary.

because the physical circulation of estuaries have been long studied. Indeed, physical models of estuarine dynamics are now found widely in engineering applications, from the construction of levies and dikes to the dredging of navigation channels. With this proliferation, ecosystem modelers can make use of well-tested, versatile physical descriptions to quantify transport and mixing rates, and turn their attention to trophic dynamics, nutrient cycling, primary production and hypoxia [e.g. Soetaert et al., 1994; Hansen et al., 1995]. As these models gain credibility, their developers are coming to the stage where they can give management advice [e.g. Smith and Hollibaugh, 1989].

1.5. Fjords

Fjords are generally distinctive systems in both physical and biological terms. They are deep, enclosed basins, usually with a restrictive sill at their seaward approaches. This relatively simple form is attractive for modeling as the demarcation is well defined. One of the few successful empirical modeling studies of eutrophication in coastal seas was provided by Gowen et al. [1992] in examining Scottish sealochs (fjords). They essentially follow the classical empirical approach of Vollenweider [1968, 1976] for fresh-water lakes. However, taking nitrogen as the limiting nutri-

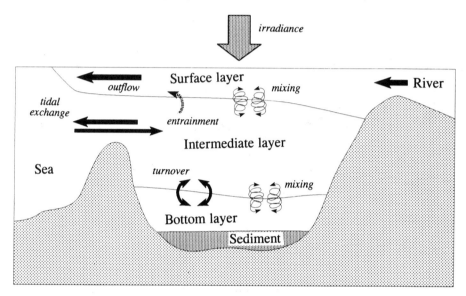

Figure 11.4. A schematic diagram of the physical features of a fjord.

ent, their method stresses the dynamic (ratio of relative change) rather than the static (ratio of concentration) relationship used in limnic systems. This is in recognition of the rapid flushing rate of marine systems compared to lakes. In this way, by analyzing 60 observation data sets, Gowen et al. [1992] estimated the yield of chlorophyll from nitrate for 21 sealochs. Further they suggested that an appropriate yield value can be used to predict the potential maximum increase in phytoplankton resulting from anthropogenic nitrogen discharge.

Given the long time scale needed to identify trends, marine scientists are not content to leave such important environmental questions to empirical formulations, and have turned their attention to dynamic modeling of fjords. Physically speaking, most fjords are estuaries being the site of the mixing of fresh and salt waters. However, because of their depth, they have an accentuated vertical structure rather then the longitudinal structure exhibited by well-mixed estuaries. Most fjords are stratified with a near-surface pycnocline due to fresh-water runoff and occasionally a deeper pycnocline due to tidal mixing (Figure 11.4). This long-lived vertical structure partitions the water column into three layers within which vertical mixing is much faster than mixing between layers. There is a slow, but measurable mixing of fresh surface water with intermediate water originating from the open sea. The bottom layer is generally quite stable. However, slow but finite mixing with intermediate-layer waters slowly reduce its density, occasionally allowing the intrusion of denser saline waters from the open sea. This causes a rapid turnover; a breakdown of the lower pycnocline and a complete mixing of the bottom and intermediate waters. Such turnover events occur perhaps once or twice a year.

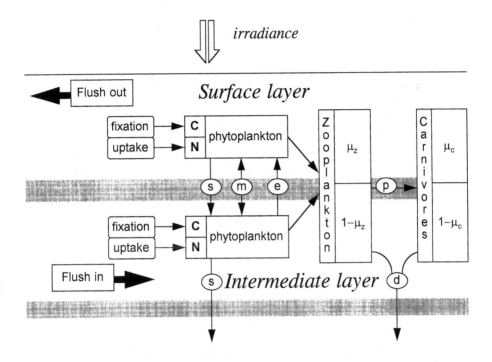

Figure 11.5. Schematic trophic interactions in a fjord as modeled by Ross et al. [1993].

Several ecosystem models of fjords within this physical context have appeared over the years [e.g. Tett, 1986; Aksnes and Lie, 1990; Rippeth et al., 1996]. As an illustrative example, we discuss here the model developed by Ross et al. [1993]. This model was designed for strategic purposes and concentrated on only a small number of dynamic interactions. In particular, biota were divided into three major trophic levels; phytoplankton, herbivores and carnivores (Figure 11.5). While this is far from a state-of-the-art description of biological interactions, ignoring e.g. species composition, competition and succession, foraging behavior, and the microbial loop, it does underline the spirit of simplicity that valuable models should embrace. These model biota were given appropriate motile behavior. Herbivorous zooplankton and carnivores could move freely between surface and intermediate layers; however, phytoplankton could only be transported vertically by water motion (mixing or entrainment) or by sinking. Nitrogen, assumed to be the limiting nutrient, cycled rapidly in the surface and intermediate layers between dissolved inorganic nitrogen and phytoplankton through uptake and excretion, and zooplankton and carnivores through excretion (Figure 11.6). The physical exchange of water between layers was derived from conservation of salt and volume, observed structure, river flow and tidal exchange, and was expressed as entrainment and mixing rates with each layer.

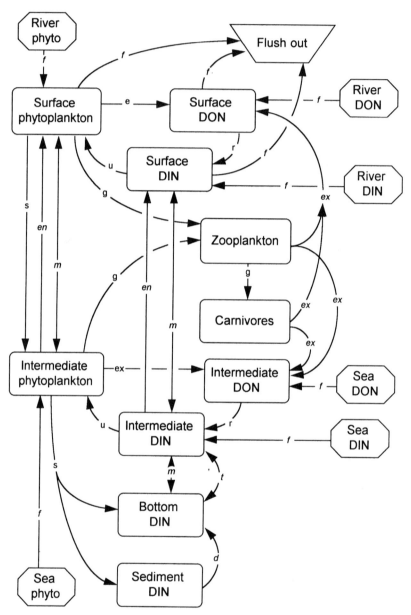

Figure 11.6. Idealized nitrogen cycling in a fjord ecosystem as incorporated in the model of Ross et al. [1993]. Cycling between the 10 nitrogen state variables (rounded boxes) are by *ex*: excretion, *r*: demineralization, u: uptake, g: grazing, s: sinking, *en*: entrainment, *m*: mixing, *t*: turnover, *d*: diffusion. Further, six external nitrogen sources (hexagonal boxes) are identified as well as a flushing out sink. Input and output of nitrogen to and from the system are by specified flow rates, *f*.

Even this simple rendering of the natural system results in a rather complex mathematical model involving:

- 14 coupled differential equations describing the 14 state variables; 10 for nitrogen cycling and four for carbon. Each of these equations is of the general form

$$dC_n/dt = a_0 C_n + \sum_{m \neq n} b_{nm} C_m + C_n \sum_{m \neq n} c_{nm} C_m + \sum_j F_j(t) \qquad (10)$$

  a slight variation of the Volterra equation, where C_n is the concentration of the nth state variable, a_0 is its growth/decay rate (e.g. excretion rate of nitrogen by phytoplankton), b_{nm} represents linear coupling to other pools, largely physical exchange processes such as the entrainment rate of intermediate dissolved inorganic nitrogen (DIN) into the surface layer, c_{nm} non-linear coupling, primarily trophic interaction (e.g. nitrogen lost from surface phytoplankton pool by zooplankton grazing), and $F_j(t)$ is external forcing functions such as the concentration of DIN in inflowing river water.

- A total of 38 parameters governing the system dynamics; 15 for phytoplankton, 10 each for zooplankton and carnivores, and 3 for nutrients. These include uptake rates, excretion rates, mortality, grazing rates etc.

- 14 external driving functions, including the seasonal variation of light intensity, nutrient concentration in river and sea water, the immigration of zooplankton and carnivores, and specification of the intensity and timing of turnover events.

The model was applied to Killary Harbour, a fjord on the west coast of Ireland, and a notional Scottish sealoch approximating the inner basin of Loch Linnhe. One surprising result was that nutrient loading was primarily driven by nutrient concentrations in the open sea rather than in the river. Internal cycling of nutrients appeared to have little influence. This was evidenced in the response of the system to turnover events. While such events increased the nutrient concentration in the euphotic zone, they were flushed out of the system faster than they could be taken up. Inorganic nitrogen supply exceeded biological demand for all but a short period at peak spring bloom. The strongest control on the size of the spring bloom was physical factors, temperature and irradiation, as well as the dynamics of the subsequent zooplankton bloom. This latter indicates a top–down control, depending ultimately on the success of zooplankton overwintering in the deep layer and/or the import of zooplankton from the adjacent sea.

In a later study [Ross et al., 1994], the model was further refined especially with respect to the grazing description and was tested for its generality on four Irish and Scottish sealochs, each with rapid flushing rates. These simulations confirmed the importance of top–down control and a further simplification was made to only three state variables: phytoplankton, zooplankton and carnivore carbon in a single layer. The fit of this minimal model was indistinguishable from that achieved by the more sophisticated dynamic description of Figure 11.5. This highlights an interesting point; that 'improving' a model does not necessarily mean increasing its level of complexity.

1.6. Coastal Seas

Of the aquatic systems mentioned here, quantifying and qualifying eutrophication in coastal seas poses the greatest challenge. Caught as they are, between the land and the deep blue sea they feel both the impact of anthropogenic fluxes and the rhythmic cycles of climatic variation played out over the deep ocean basins. That there are natural, interannual variations in ocean – atmosphere coupling is well known, El Niño and the North Atlantic oscillation for instance. Although not rigorously demonstrated, it is likely that these variations impact shelf seas also; the 'great salinity anomaly' of the 1970s in the North Sea for instance [e.g. Turrell, 1992; Lindeboom et al., 1994]. While the term eutrophication is neutral saying nothing about the source of the cause or the quality of the effect, for policy makers there is a clear distinction: there are those undesirable effects that can be regulated for, and those that cannot. Clearly, a proper understanding of environmental dynamics is crucial for sound policy implementation.

Given the complex dynamics played out in coastal seas, empirical modeling of coastal eutrophication is rare. One notable exception is Nixon [1992] who reports the relationship between dissolved inorganic nitrogen (DIN) for primary production and phytoplankton biomass in a wide range of marine environments. It was found that at loading less than 5 mmol m^{-2} yr^{-1} the net primary production was less than that which would be expected by the Redfield ratio indicating the significance of nitrogen recycling at low input rates. The rigor of this relationship is yet to be fully explored.

Eutrophication by nutrient loading is a two-pronged problem. Firstly, there is the question of how these nutrients are dispersed from a source, and secondly, their subsequent stimulation of primary production. In this, the distribution and availability of nutrients in coastal seas are intimately connected with buoyancy effects (see chapter 3). This is exhibited in two phenomena; nutrients derived from riverine sources carried in buoyant low salinity coastal plumes and the stratification and destratification of the seasonal thermocline and its role in governing the vertical cycling of nutrients.

An increased understanding of the coupling between physical and biological processes has resulted in the hypothesis that the seasonal cycles of phytoplankton in shelf seas can be explained in terms of the relationship between vertical mixing and the requirement of planktonic microalgae for light and nutrients [Pingree et al., 1977]. The key to this system is the development of the seasonal thermocline; the period when light conditions are suitable for photosynthesis coincide with developing water column stability so that nutrients lost form the euphotic zone tend to remain lost for the duration. Conditions suitable for plankton blooms only occur for a short period in the spring and occasionally in the autumn; for the most part conditions for phytoplankton growth are relatively poor because of inadequate light in the winter and limiting nutrients in the summer.

While this is the seasonal scenario, a similar mechanism is thought to operate for episodic nuisance blooms often occurring in summer. Here a combination of strong winds and/or a spring tide temporally reduces water column stability, injecting nutrients into the euphotic zone, and triggering the bloom.

Over the years a number of mathematical models formulated to investigate this system has appeared [e.g. Radach and Maier-Reimer, 1975; Jamart et al., 1977; Steele and Henderson, 1981; Fasham and Platt, 1983; Taylor et al., 1986; Tett et al., 1986; Stigebrandt and Wulff, 1987], each with varying levels of sophistication and success. One distinctive difference between these was how vertical mixing processes were treated. These ranged from a uniform diffusivity, to entrainment into the surface wind-mixed layer, to a state-of-the-art dynamic description.

An example of a vertical 'point' model that serves to illustrate this is that formulated by Sharples and Tett [1994]. In this and other such models, the advection of nutrients and lateral buoyancy effects are considered unimportant and not modeled, thus restricting their suitability to, for instance, the central gyre of the North Sea. None the less, it accentuate some important processes with potential application to eutrophication. The model formulation can be divided into two aspects as follows:

Physical:

- Momentum is imparted to the water column by a tidally varying pressure gradient and surface wind stress and lost through friction at the bottom. The vertical stress is given by $\tau(z) = N_z\, \partial u/\partial z$ where u is the horizontal velocity vector and N_z is the vertical turbulent viscosity. Thus

$$u/\partial t + f z \times u - \partial (N_z\, \partial u/\partial z)/\partial z = -A \sin(\omega t) \qquad (11)$$

  with surface and bottom boundary conditions respectively $\tau_s = k_s |w| w$ and $\tau_b = k_b |u| u$.

- Density changes within the water column are brought about by surface heating and cooling and turbulent vertical diffusion. Vertical mixing of temperature T is regulated by

$$T/\partial t - \partial (K_z\, \partial T/\partial z)/\partial z = 0 \qquad (12)$$

  where K_z is the vertical turbulent diffusivity.

- Turbulent properties N_z and K_z are calculated dynamically through a level-2 turbulent closure scheme [Mellor and Yamada, 1982]. Essentially this scheme recognizes that the vertical mixing of momentum and material becomes suppressed as the water column stratifies. Competing with this is the turbulent cascade of energy from the macro- to the micro-scale; energy which is used in part to erode stratification. Formulations of this scheme are many and varied and will not be reproduced here. Interested readers can refer to Luyten et al. [1996], one of the more readable treatments of the subject.

Biological:

Four depth- and time-dependent state variables were considered in describing biological interactions:

- Phytoplankton biomass X: $\partial X/\partial z = \partial (K_z \partial X/\partial z)/\partial z + \mu X - gX$ $\hspace{1em}$ (13)

  μ is the reproduction rate and g the grazing rate by zooplankton

- Dissolved inorganic nitrogen S: $\partial S/\partial t = \partial (K_z \partial S/\partial z)/\partial z - uX + egX$ $\hspace{1em}$ (14)

  u is the uptake rate by phytoplankton and e is the fraction excreted by zooplankton.

- Internal algal nutrient N: $\partial N/\partial t = \partial (K_z \partial N/\partial z)/\partial z + uX - gN$ $\hspace{1em}$ (15)

- PAR (photosynthetically active radiation) I: $\partial I/\partial z = - I (\lambda_0 + \epsilon X)$ $\hspace{1em}$ (16)

  where ϵ simulates the self shading by phytoplankton, and λ_0 is the attenuation coefficient.

One novel feature of the model is that it includes a biological influence on the physics, namely the heating and stratification of the water column by absorption of solar radiation Q_s by phytoplankton X. Specifically

$$\partial Q_s / \partial z = - Q_s (\lambda_0 + \epsilon X)$$

The model was used to explore mechanisms involved in sustaining the subsurface chlorophyll maximum observed in the central North Sea. The potential importance to eutrophication is discussed in chapter 5. With respect to this, three hypotheses that have been presented were tested:

1. Pingree et al. [1977] suggested that a slow leakage of nutrients into the thermocline coupled to the relatively long residence time of phytoplankton in that region can lead to enhanced productivity.
2. Steele and Yentsch [1960] suggested that phytoplankton sinking rates decrease as nutrient supplies increase allowing them to take advantage of higher ambient nutrient levels near the thermocline.
3. Margalef [1978] and Tett [1987] proposed that dinoflagellates and diatoms can migrate vertically to take up nutrients at depth and photosynthesis near the surface.

Of the model simulations incorporating these mechanisms, only (1) produced a subsurface chlorophyll maximum (Figure 11.7): that is it is due to purely physical effects, the slow diffusive (background) entrainment of nutrients into a stabilized layer of the water column rather than a behavioral adaptation of the phytoplankton.

A second effect explored by this model was how it responded to varying detail in the wind forcing which is the major source of vertical mixing in the central North Sea. In particular, a comparison was made between model results driven by seasonally averaged winds, and more detailed wind forcing from daily observations. It was found that the more realistic meteorological inputs reproduced the subsurface

Height, m

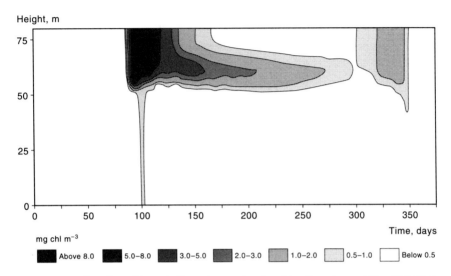

Figure 11.7. Subsurface chlorophyll maximum simulated by Sharples and Tett [1994], for a notional central North Sea situation. Forcing is simulated by monthly mean meteorological conditions (wind, temperature, light), and a gentle entrainment of nutrients into the thermocline is included, testing the hypothesis postulated by Pingree et al. [1977].

maximum without recourse to the artificially induced background diffusion used in testing hypothesis (1) above, thus identifying the source of the slow leakage of nutrients into the euphotic zone, and highlighting the need for sound physical frameworks upon which to develop ecological models.

While this class of model has found its widest application to date as a research tool, some investigators are turning their attention to eutrophication and its associated influence. An indication of this was presented by Stigebrandt and Wulff [1987] who developed a time-dependent physical-bio-geochemical model with high vertical resolution to investigate nutrient and oxygen conditions in the Baltic Sea. While this model had poor horizontal resolution, it showed reasonable correlation with 20 years of observations in the East Gotland Basin.

Closer to the coast, ecological dynamics are influenced as much by horizontal as by vertical transport and mixing. These complex systems are only now drawing the attention required to model their physical aspects [e.g. Simpson et al., 1993]. Given this, fully interactive physical and biological models of such systems are as yet not forthcoming. Nonetheless, some simplified dynamic descriptions have shown some success in investigating eutrophication. For instance, it is estimated that the nitrogen and phosphorus load carried by the Rhine into the southern Bight of the North Sea has increased 6-fold since the 1930s. Especially in the coastal areas, silicate is now the first nutrient depleted by spring blooms as opposed to nitrogen and phosphorus 60 years ago. To what extent this trend has given rise to enhanced phytoplankton growth and changes in species composition was explored by Fransz and

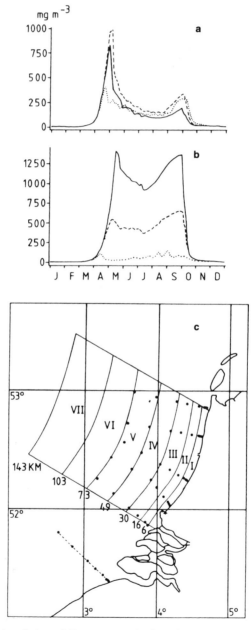

Figure 11.8. Seasonal variation of diatom (a) and other phytoplankton (b) carbon within a 6 km wide strip of the Netherlands coast (strip I of c) as simulated by Fransz and Verhagen [1985]. Simulations are for (a) 1980 conditions, (b) 1930 conditions and (c) 1930 river discharge with 1980 ambient sea concentrations. Similar variations, although smaller in intensity are exhibited in other model strips extending beyond 100 km from the shore.

Verhagen [1985] using a dynamic model. While they were cautious in ascribing too much significance to their results (there being a paucity of observational evidence for validation) they showed, as illustrated in Figure 11.8:

- Diatom biomass has not altered appreciably over the last 60 years. While there is some indication of change in the near shore zone (Figure 11.8), 2-fold increase in the spring bloom and a 2-fold decrease in the autumn bloom, during the summer months, and indeed integrated over the whole year, the diatom biomass has remained relatively unchanged.
- In contrast, the biomass of other phytoplankton species, mainly dinoflagellates, has increased by a factor of 2 to 4 between 1930 and 1980. Half of this increase appears to be due to the increased loading of N and P in the Rhine, and the other half from increased concentrations in the coastal sea, also from anthropogenic sources discharged through other rivers, e.g. Meuse and Scheld.
- 60 years ago, diatoms dominated the phytoplankton of the Dutch coastal zone. Now, this has switched to other, non silicate dependent species.

These types of calculation, despite their high degree of uncertainty, direct researchers in their planning of future studies, both in identifying possible processes and in the design of observational programs needed to corroborate their importance.

With the advent of cheap, readily available and massive computing power, a significant section of the oceanographic community has embarked on the numerical modeling of large-scale ocean systems. This trend, first evident in physical oceanography, is becoming more popular in the study of large-scale, shelf-sea ecosystems. An extensive review of these types of models as they pertain to the North Sea was provided by Fransz et al. [1991]. More recently, fully developed, 3-dimensional coupled physical-biological models are becoming available, e.g. Aksnes et al. [1995] for the North Sea, Tamsalu and Myrberg [1995] for the physical component and Tamsalu and Ennet [1995] for the biological component of a Baltic ecosystem model.

1.2. Conclusions

Eutrophication studies of lakes are at the stage where predictive models are widespread and successful. This is not so for marine systems. The single-most complicating factor in marine ecosystems is that they are unbounded. Length scales in marine systems are rarely geographical, but depend on governing processes often in combination; e.g. an advective velocity times a mechanism's time scale. A proper description of nutrients, their sources and cycling together with water motion and mixing is a fundamental criterion in understanding and regulating for such potentially catastrophic events as massive or toxic plankton blooms and anoxia. Ecological and marine scientists are meeting the challenges posed by these questions in the many and varied environments in which they are found.

Empirical modeling has its place in illuminating relationships and accentuating processes. However, it is becoming clear that coupled physical-biological models represent the best avenue toward the future. This is in recognition that eutrophication is not simply a question of nutrient loading, but that the pathways through which nutrients impact on marine productivity are many and varied, being governed as much by physical as biological processes. Here much can be learned from the large body of expertise existing in the physical and ecological oceanographic modeling community. While the concept of slipping biological modules into existing physical models is appealing, it may in practice prove far from straightforward.

In the coming years, managers and policy makers may expect marine models to come closer to realizing their potential as predictive tools. In this, it is anticipated that decision makers will work closely with model developers, in particular

- in defining the question to be addressed
- helping to establish adequate monitoring programs, the cost effectiveness of which can be gauged by sensitivity analysis
- applying risk assessment to establish reasonable model validation criteria.

Finally, it should be noted that an ecological model, no matter how well tested and validated, can never be a substitute for monitoring and scientific interpretation. Their implementation and interpretation demand a 'heads up' approach. Models cannot be treated as off-the-shelf products which can be blindly applied to provide answers; the scientific community as well as decision makers are ill served by those who suggest they can. Nonetheless, the quantitative and predictive power of models is hard to ignore, especially in these days of unprecedented and rapid environmental change.

References

Ahlgren, I., T. Frisk, and L. Kamp-Nielsen, Empirical and theoretical models of phosphorus loading, retention and concentration vs. trophic state, *Hydrobiol., 170*, 285–303, 1988.

Aksnes, D. L., and U. Lie, A coupled physical-biological pelagic model of a shallow sill fjord, *Estuar. Coast. Shelf Sci., 31*, 459–486, 1990.

Aksnes, D. L., K. B. Ulvestad, B. M. Balino, J. Berntsen, J. K. Egge, and E. Svendsen, Ecological modeling in coastal waters: towards predictive physical-chemical-biological simulation models, *Ophelia, 41*, 5–36, 1995.

Bowden, K. F., Turbulence and mixing in estuaries, in *The Estuary as a Filter* edited by V. S. Kennedy, pp. 15–26, Academic Press, New York, 1984.

Fasham, M. J. R., and T. Platt, Photosynthetic response of phytoplankton to light: a physiological model, *Proc. R. Soc. Lond. Ser. B, 219*, 355–370, 1983.

Franks, P. J. S., Coupled physical-biological models in oceanography, *Rev. Geophys., Suppl.,* 1177–1187, 1995.

Fransz, H. G., and J. H. G. Verhagen, Modeling research on the production cycle of phytoplankton in the southern Bight of the North Sea in relation to riverborne nutrient loads, *Neth. J. Sea Res., 19*, 241–250, 1985.

Fransz, H. G., J. P. Mommaerts, and G. Radach, Ecological modeling of the North Sea, *Neth. J. Sea Res., 28*, 67–140, 1991.

Gowen, R. J., P. Tett, and K. J. Jones, Predicting marine eutrophication: the yield of chlorophyll from nitrogen in Scottish coastal waters, *Mar. Ecol. Prog. Ser., 85*, 153–161, 1992.

Grantham, B., and P. Tett, The nutrient status of the Clyde Sea in winter, *Estuar. Coast. Shelf Sci.*, *36*, 449–462, 1993.

Hansen, I. S., G. Ærtebjerg, and K. Richardson, A scenario analysis of effects of reduced nitrogen input on oxygen conditions in the Kattegat and Belt Sea, *Ophelia*, *42*, 75–93, 1995.

Jumart, B. M., D. F. Winter, K. Banse, G. C. Anderson, and R. K. Lam, A theoretical study of phytoplankton growth and nutrient distribution in the Pacific Ocean off the northwest US coast, *Deep-Sea Res.*, *24*, 753–773, 1977.

Kristensen, P., J. P. Jensen, and E. Jeppesen, Eutrofieringsmodeller for Søer, NPO-forskning fra Miljøstyrelsen, c 9, 120 pp., Ministry of Environment, Copenhagen, 1990.

Legendre, L., and S. Demers, Towards dynamic biological oceanography and limnology, *Can. J. Fish. Aquat. Sci.*, *41*, 2–19, 1984.

Lindeboom, H. J., W. van Raaphorst, J. J. Beukema, G. C. Cadée, and C. Swennen, (sudden) changes in the biota of the North Sea: Oceanic influences underestimated? *ICES CM 1994/L:27*, 16 pp., ICES, Copenhagen, Denmark.

Luyten, P. J., E. Deleersnijder, J. Ozer, and K. G. Ruddick, Presentation of a family of turbulence closure models for stratified shallow water flows and preliminary application to the Rhine outflow region, *Cont. Shelf Res.*, *16*, 101–130, 1996.

Margalef, R,. Life-forms of phytoplankton as survival alternatives in an unstable environment, *Oceanol. Acta*, *1*, 493–509, 1978.

Mellor, G. L., and T. Yamada, Development of a turbulence closure model for geophysical fluid problems, *Rev. Geophys. Space Phys.*, *20*, 851–875, 1982.

Nicholls, K. H., and P. J. Dillon, An evaluation of phosphorus-chlorophyll-phytoplankton relationships in lakes, *Int. Revue Ges. Hydrobiol.*, *63*, 141–154, 1978.

Nihoul, J. C. J., Applied mathematical modeling in marine science, *App. Math. Modell.*, *1*, 3–8, 1976.

Nixon, S. W., Quantifying the relationship between nitrogen input and the productivity of marine ecosystems, *Proc. Adv. Mar. Techn. Conf.*, *5*, 57–83, 1992.

Nixon, S. W., Coastal marine eutrophication: a definition, social causes and future concerns, *Ophelia*, *41*, 199–219, 1995.

Officer, C. B., Discussion of the behaviour of non-conservative dissolved constituents in estuaries, *Estuar. Coast. mar. Sci.*, *9*, 91–94, 1979.

Officer, C. B., and D. R. Lynch, Dynamics of mixing in estuaries, *Estuar. Coast. Shelf Sci.*, *12*, 525–533, 1981.

Officer, C. B., and J. H. Ryther, The possible importance of silicon in marine eutrophication, *Mar. Ecol. Prog. Ser.*, *3*, 83–91, 1980.

Pejrup, M., J. Bartholdy, and A. Jensen, Supply and exchange of water and nutrients in the Grodyb tidal area, Denmark, *Estuar. Coast. Shelf Sci.*, *36*, 221–234, 1993.

Pingree, R. D., L. Maddock, and E. I. Butler, The influence of biological activity and physical stability in determining the chemical distribution of inorganic phosphate, silicate and nitrate, *J. Mar. Biol. Ass. U.K.*, *57*, 1065–1073, 1977.

Platt, J. R., Strong interference, *Science*, *146*, 347–353, 1964.

Platt, T., K. H. Mann, and R. E. Ulanowicz, *Mathematical Models in Biological Oceanography*, Unesco Press, Paris, 156 pp. 1981.

Radach, G., and E. Maier-Reimer, The vertical structure of phytoplankton growth dynamics. A mathematical model, *Mem. Soc. R. Sci. Liege*, *7*, 113–146, 1975.

Rippeth, T. P., and K. J. Jones, The seasonal cycle of nitrate in the Clyde Sea, *J. Mar. Syst.*, in press, 1996.

Ross, A. H., W. S. C. Gurney, M. R. Heath, S. J. Hay, and E. W. Henderson, A strategic simulation model of a fjord ecosystem, *Limnol. Oceanogr.*, *38*, 128–153, 1993.

Ross, A. H., W. S. C. Gurney, and M. R. Heath, A comparative study of the ecosystem dynamics of four fjords, *Limnol. Oceanogr., 39*, 318–343, 1994.

Sharples, J., and P. Tett, Modeling the effect of physical variability on the midwater chlorophyll maximum, *J. Mar. Res., 52*, 219–238, 1994.

Scheffer, M., and J. Beets, Ecological models and the pitfalls of causality, *Hydrobiol. 275/276*, 115–125, 1994.

Simpson, J. H., W. G. Bos, F. Schirner, A. J. Souza, T. Rippeth, S. E. Jones, and D. Hydes, Periodic stratification in the Rhine Rofi in the North Sea, *Oceanol. Acta, 16*, 23–32, 1993.

Smith, S. V. and J. T. Hollibaugh, Carbon-controlled nitrogen cycling in a marine 'macrocosm': an ecosystem-scale model for managing cultural eutrophication, *Mar. Ecol. Prog. Ser., 52*, 103–109, 1989.

Smith, S. V., and J. T. Hollibaugh, Coastal metabolism and oceanic organic carbon balance, *Rev. Geophys., 31*, 75–89, 1993.

Smith, S. V., and H. H. Veeh, Mass balance of biochemically active materials (C, N, P) in a hypersaline gulf, *Estuar. Coast. Shelf Sci., 29*, 195–215.

Soetaert, K., P. M. J. Herman, and J. Kromkamp, Living in the twilight: estimating net phytoplankton growth in the Westerschelde estuary (the Netherlands) by means of an ecosystem model (MOSES), *J. Plankton Res., 16*, 1277–1301, 1994.

Steele, J. H., and E. W. Henderson, A simple plankton model, *Am. Nat., 117*, 676–691, 1981.

Steele, J. H., and C. S. Yentsch, The vertical distribution of chlorophyll, *J. Mar. Biol. Ass. U.K., 39*, 217–226, 1960.

Stigebrandt, A., and F. Wulff, A model for the dynamics of nutrient and oxygen in the Baltic proper, *J. Mar. Res., 45*, 729–759, 1987.

Tamsalu, R., and K. Myrberg, Ecosystem modeling of the Gulf of Finland. I. General features and the hydrodynamic prognostic model FINEST, *Estuar. Coast. Shelf. Sci., 41*, 249–273, 1995.

Tamsalu, R., and P. Ennet, Ecosystem modeling of the Gulf of Finland. II. The aquatic ecosystem model FINEST, *Estuar. Coast. Shelf. Sci., 41*, 429–458, 1995.

Taylor, A. H., J. R. W. Harris, and J. Aiken, The interaction of physical and biological processes in a model of the vertical distribution of phytoplankton under stratification, in *Marine Interfaces Hydrodynamics*, edited by J. C. J. Nihoul, pp. 313–330, Elsevier Oceanography Series, 42, Amsterdam, 1986.

Tett, P., Modeling phytoplankton production at shelf sea fronts, *Phil. Trans. R. Soc. Lond. A, 302*, 605–615, 1981.

Tett, P., Modeling the growth and distribution of marine microplankton, in *Ecology of Microbial Communities*, pp. 387–425, Cambridge University Press, 1987.

Tett, P., A. Edwards, and K. Jones, A model for the growth of shelf sea phytoplankton in summer, *Estuar. Coast. Shelf Sci., 23*, 641–672, 1986.

Turrell, W. R., New hypothesis concerning the circulation of the northern North Sea and its relation to North Sea fish stock recruitment, *ICES J. Mar. Sci., 49*, 107–123, 1992.

van Raaphorst, W., P. Ruarij, and A. G. Brinkman, The assessment of benthic phosphorus regeneration in an estuarine ecosystem model, *Neth. J. Sea Res., 22*, 23–36, 1988.

Vollenweider, R. A., Water Management Research, OECD, Paris DAS/CSI, 68.27, 1968.

Vollenweider, R. A., Advances in defining critical loading levels for phosphorus in lake eutrophication, *Mem. Ist. It. Idrobiol., 33*, 53–83, 1976.

Vollenweider, R. A., and J. Kerekes, Eutrophication of Waters, Monitoring, Assessment and Control, OECD, Paris, 1982.

12

Conclusion, Research and Eutrophication Control

Katherine Richardson

12.1. Introduction

From the preceding chapters, it is clear that eutrophication can have profound effects on coastal marine ecosystems. However, the effect of nutrient enrichment on nutrient metabolism in marine environments is a complicated phenomenon that will depend upon the quantities and species of nutrients being delivered to the system as well as the mechanism(s) and timing of the delivery. In addition, the hydrographic, geological (i.e. bottom type) and biological characteristics of the recipient are important in determining the effect that eutrophication will have on an ecosystem. Thus, it is difficult to generalize about the effects of eutrophication in marine coastal environments and prediction of the potential effects in any given environment will require thorough analysis of the physical, chemical and biological processes occurring there.

12.2. Nutrient Limitation, Sources and Fluxes

From available information, there is no reason to believe that the initial biological response to eutrophication in temperate regions differs greatly between fresh and marine waters or from one marine region to another. Plant biomass in all of these environments is, at least during some periods, limited by nutrient availability and increasing nutrient input via eutrophication will, when nutrient limitation is in effect, increase plant growth and/or change the species composition of the plant community.

In most fresh waters, phosphorus (P) availability is accepted to be limiting for plant growth. Thus, most remedial actions with respect to eutrophication in fresh water have focused on P control and attempts to describe the regulation of productivity in

Eutrophication in Coastal Marine Ecosystems
Coastal and Estuarine Studies, Volume 52, Pages 243–267
Copyright 1996 by the American Geophysical Union

these systems have generally been based on mass balances with P (see chapter 11). In marine systems, however, the situation is more complicated. Here, it is generally argued that nitrogen (N) rather than P is usually the potentially limiting nutrient for primary production [e.g., Smetacek et al., 1991].

This argument is not universally accepted (see chapter 1, section 1.4). However, it does seem clear, on the basis of a comparison of the elemental composition of phytoplankton and typical fresh and marine waters, that nitrogen, phosphorus and silica are all potential candidates for limiting plant growth in marine waters. In fresh waters, phosphorus is the macronutrient most likely to limit phytoplankton production (see chapter 1, this volume, and Hecky and Kilham [1988]). The possibility that N and/or other nutrients may be limiting for plant biomass in marine systems makes it much more difficult to establish mass balance descriptions of such regions.

This is because there appears to be the possibility for different nutrients to be limiting for phytoplankton growth at different times in marine waters. Thus, it is not immediately obvious which nutrient should be chosen for the establishment of a mass balance to describe production in marine systems. However, even if, as many suggest, nitrogen availability controls production in most marine systems, establishing an N mass balance is not as straight forward as establishing a P mass balance. This is because biochemical processes (denitrification and nitrogen fixation) in addition to geochemical processes (input/removal/burial) control nitrogen availability in aquatic ecosystems. Similar biochemical processes with respect to phosphorus input/removal are not found. Reliable data on the rates for the input/removal of nitrogen via biochemical processes are still scarce.

In addition, atmospheric delivery of nitrogen to the coastal oceans can be a quantitatively important input process. This is not the case for P input. Atmospheric N contributions to total N delivery in most marine systems are, however, not yet well quantified. The greater number of processes affecting N input/output to coastal marine systems make the establishment of a quantitative mass balance for N more difficult than for P. Thus, many of the advances made in the field of limnology with respect to modeling eutrophication effects and the potential environmental improvement that might be expected following remedial actions are not directly applicable to coastal marine environments.

12.3. Establishing Cause and Effect Relationships

It is very difficult to quantify and, thus, predict causal relationships when a change in ecosystem structure due to eutrophication is observed in the marine environment. Unequivocal identification of cause and effect usually requires that the "effect" in question can be reproduced in a test system when the system is subjected to the suspected causal treatment. Manipulation of entire marine systems is usually not possible. "Mesocosm" (small-scale experimental model of the study system) studies have been valuable in clarifying many of the ecosystem responses observed in the wake of marine eutrophication [e.g., Isaksson et al., 1994; Hinga, 1990; Oviatt et al., 1989].

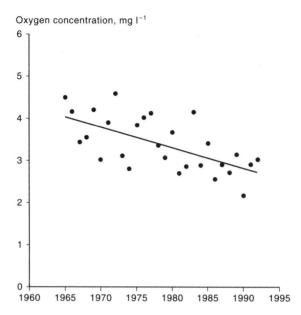

Figure 12.1. Average annual oxygen concentrations (mg l⁻¹) during the pooled months of August, September and October in bottom water (deepest available measurement below 19 m) in the region enclosed by the following coordinates: 55°55'–56°40'N; 10°40'–12°30'E. Data from the International Exploration of the Seas (ICES) Oceanographic Database.

However, there are some fundamental differences between mesocosms and the systems they are designed to mimic. The differences usually identified as being the most important are those associated with the artificial boundaries imposed on the system by the mesocosm (i.e. lack of or only limited contact to natural sediment, growth of microorganisms on the walls of the mesocosm, etc.). It is often argued that these differences in boundary conditions may alter processes related to nutrient turnover with respect to those occurring in nature. Thus, it is not all questions relating to cause and effect in ecosystem responses to eutrophication that can be answered with the help of a mesocosm.

In natural marine systems, it is most common that cause and effect with respect to eutrophication responses are argued on the basis of correlations between different changes/responses occurring within the system. This statistical approach, however, merely indicates that two variables change in a consistent manner in relation to each other. Such an approach does not identify cause and effect relationships with certainty. As a result, there is often considerable public and scientific debate concerning whether cause and effect relationships in response to eutrophication are established with a degree of certainty that warrants specific remedial actions.

In the Kattegat, for example, it is well established that oxygen concentrations during the late summer months have been decreasing during the period 1965–1992 (Figure 12.1). The commonly used ¹⁴C method of measuring aquatic primary production was developed by Steemann Nielsen [1952] who carried out much of his early work using the method in the Kattegat. As a result, there exists an unusually long time series of data on primary production using sensitive radioactive methodology for this marine region. This time series suggests that there has been an increase in the phytoplankton primary production occurring here in the period between the 1950s and the late 1980s–early 1990s (see chapter 5, this volume, and Richardson and Heilmann [1995]). This increase is difficult to quantify owing to differences in the methods used during the two study periods. However, Heilmann and Richardson argue that pelagic primary production has increased by at least a factor of two between the two study periods.

Thus, the reduction in oxygen concentrations observed in the bottom waters of the Kattegat during the last thirty years correlates with an increase in pelagic primary production occurring in the same period. While this correlation does not necessarily indicate cause and effect, there is strong evidence from mesocosms and some natural (especially fresh-water) systems that an increase in the magnitude of primary production results in an increase in hypoxia in bottom waters (see chapter 1 and references therein).

Accepting there is a likely causal link between increased pelagic primary production and increased hypoxia, we then would like to determine the cause of the increased primary production. As identified earlier (chapter 1), primary production is generally limited by light or nutrient availability. There is no evidence of a consistent change in the light climate in the Kattegat that would stimulate primary production during the period of interest (S.E. Jensen, The Royal Veterinary and Agricultural University, Copenhagen: unpublished data). Thus, it would appear that the most likely cause of the increased primary production is an increase in nutrient availability.

Nutrient availability is, however, affected by both natural and anthropogenic processes. The hydrographic delivery of nutrients into and out of the photic zone determines the magnitude of primary production (see chapters 4 and 5). Until recently, most interest in terms of eutrophication's influence on phytoplankton productivity has been directed toward advective processes (the horizontal delivery and removal of material) affecting nutrient availability. However, the vertical delivery, via atmospheric deposition and upwelling of nutrients from bottom waters to the surface layer is also critical to the production processes occurring in stratified coastal waters. Examples of the importance of vertical nutrient fluxes for pelagic processes in the Kattegat can be found in chapters 2, 4, and 5.

Upwelling is associated with a weakening or breakdown of stratification. Such changes are ultimately dependent upon weather conditions: wind (direction and strength) and precipitation (in the case of the Kattegat, for example, fresh-water input to the Baltic is important in determining surface salinity). The exact relationship between changes in nutrient availability affecting primary production and weather conditions is still not quantified. However, it may be predicted that a greater production of turbulent kinetic energy during the summer months (i.e. prior to the development of hypoxia in bottom waters and when stratification of the

water column is greatest) will enhance upwelling of nutrients from bottom into surface waters and stimulate phytoplankton growth. At the same time, turbulence in
the surface layer may promote aggregate formation (see chapter 4). As large aggregates sink more quickly than smaller particles, such turbulence events may result in
an increase in the delivery of organic material to the bottom. It may, then, be argued that an increase in turbulence during summer months may stimulate hypoxia
by increasing the phytoplankton production and by inducing delivery of organic
material to the bottom.

On the other hand, turbulence will also stimulate the delivery of oxygen to bottom
waters which will reduce the likelihood of a serious hypoxia event occurring later
during the season. Thus, it is not clear how a long-term change in the production of
turbulent kinetic energy during summer months when the water column is permanently stratified might be expected to affect the probability and intensity of hypoxia
events. Turbulence generated during the late-summer/autumn months when hypoxia is occurring, however, will have a mitigating effect on hypoxia events by increasing oxygen delivery to bottom waters.

Jacobsen [1994] examined long-term (1890–1988) variations in the production of
turbulent kinetic energy from wind and currents in the Great Belt area of the
Kattegat. For most of the Kattegat, upwelling of nutrients is a function of wind
generated turbulence. Here, the wind speed cubed (W^3) can be used as a proxy for
wind generated turbulence. Jacobsen has calculated W^3 for the different seasons of
the year (winter = December, January, February, spring = March, April, May and
so on) for three different stations in the Great Belt (Figure 12.2).

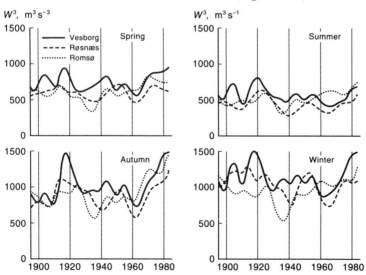

Figure 12.2. Cubic velocity of wind speed (W^3) at three different stations in the Kattegat. The
coordinates of the stations are (Vesborg: 55°46'N 10°33'E; Røsnæs: 55°45'N 10°52'E;
Romsø: 55°31N 10°48'E) as a function of year). W^3 is used here as a proxy for turbulent kinetic energy added to the water column from wind activity. From Jacobsen [1994].

From this figure, it can be seen that there appears to have been an increase in W^3 during autumn and winter since about 1960. The pattern is less convincing for the spring and summer months. The lack of a clear pattern in the wind conditions during summer months makes it unlikely that a long-term change in wind conditions during the period can explain the observed increase in primary production and hypoxia events in this area. On the contrary, it may be argued that the apparent increase in wind generated turbulence that has occurred during the autumn months should have reduced the intensity of hypoxia events owing to its having increased the delivery of oxygen to bottom waters during the period when hypoxia is most likely to occur.

From the available information, it is not possible to totally eliminate the possibility that some natural change or combination of natural changes in the forcing conditions for nutrients in the photic zone may be related to the changes observed in pelagic primary production. However, with our current understanding of the dynamics of marine ecosystems, it seems most likely that the increases in the frequency and intensity of hypoxia events in the Kattegat are a result of an increase in the availability of nutrients which has stimulated the production of organic material originating through phytoplankton primary production.

12.4. Long-term Changes in Anthropogenic Nutrient Delivery to the Kattegat

There are data showing an increase in anthropogenic input of nutrients to the Kattegat since the 1950s. Hansen et al. [1995] have argued that anthropogenic nitrogen input to these waters doubled between the 1950s and the 1980s (see chapter 3). Data do not appear to be available in order to conduct similar analyses for phosphorus [Hansen et al., 1995].

While the increase in anthropogenic input of nitrogen to the Kattegat may have increased by a factor of two, the absolute magnitude of the nutrient sources most affected by anthropogenic activities (i.e. runoff and atmospheric input) is still relatively small (214 thousand tonnes per annum) in terms of the gross nitrogen delivery to the Kattegat (794 tonnes per annum from Table 12.1). This discrepancy between the size of the local anthropogenic input of nutrients to the region and the total nutrient delivery has led to considerable discussion as to whether or not remedial programs designed to reduce eutrophication from local sources can be expected to result in a measurable improvement in oxygen conditions in the Kattegat.

There are, however, several important factors that need to be taken into account when considering eutrophication effects and their relationship to the local anthropogenic nutrient input versus more distant nutrient sources. The first is that it is *net* rather than *gross* delivery of nutrients that is most relevant in terms of the long-term response of the ecosystem. Large amounts of nutrients exit the Kattegat over both the Baltic and (especially) the Skagerrak borders. Thus, the net nutrient delivery over these borders is considerably less than the gross delivery. As runoff and atmospheric delivery of nitrogen is almost entirely net delivery, the relative importance of these sources in the net budget is much greater than in the gross N budget for the Kattegat (Table 12.1).

Relating such nitrogen budgets to phytoplankton activity is further complicated by the fact that these budgets are most often created for total nitrogen or phosphorus input. Considerable amounts of nitrogen are bound up in complex organic compounds (for example, humic substances). Most phytoplankton are not able to efficiently utilize such organic nitrogen compounds as a nutrient source. In addition, the turnover time (i.e. time required to break the compound down to its inorganic constituents) for many organic nitrogen compounds is large relative to the residence time for water masses in many areas (including the Kattegat where the average water exchange is on the order of four months [Hansen et al., 1995]). Thus, the nitrogen bound up in these organic compounds is often not immediately available for phytoplankton growth and it is theoretically possible for such compounds to be transported through the Kattegat without ever being incorporated into phytoplankton.

The percentage of organic nitrogen varies from each of the sources shown in Table 12.1. The highest percentage (>90%) is found in Baltic Sea water entering the Kattegat. Thus, of the net delivery of 122,000 tonnes of nitrogen entering the Kattegat yearly from the Baltic only an estimated 8 tonnes is estimated to be in an inorganic form immediately available for phytoplankton growth [Richardson and Ærtebjerg, 1991] (Table 12.1). Given that most of the nitrogen entering the Kattegat via local anthropogenic input is in an inorganic form that is immediately available to phytoplankton, the potential of local nitrogen input to influence phytoplankton growth in the Kattegat may actually be greater than that of Baltic Sea nitrogen even though it initially appears that more nitrogen is coming from the latter source.

Similarly, the timing of nutrient delivery is important in determining the relative importance of a particular nutrient source to phytoplankton growth. In surface

TABLE 12.1. Annual Nitrogen (N) and Phosphorus (P) input to the Kattegat in 1000 tonnes. From Richardson and Ærtebjerg [1991].

	Baltic	Skagerrak	Atmosphere	Denmark	Sweden	Germany	Total
"Gross" input total N	180	400	60	90	45	19	794
"Net" input total N	122	−173	49	90	45	19	152
"Net" input inorganic N	8	54	49	75	34	16	236

	Baltic	Skagerrak	Atmosphere	Denmark	Sweden	Germany	Total
"Gross" input total P	18	62	0.5	5.8	1.6	2.6	90.5
"Net" input total P	10.7	6.6	0.5	5.8	1.6	2.6	14.6
"Net" input inorganic P	2.2	4.7	0.5	5.0	0.6	2.0	15.0

waters, phytoplankton are apparently limited by nutrient availability from the period following the spring bloom until the autumnal breakdown of stratification (see chapter 5). Thus, nutrients entering the surface water during this time will potentially stimulate phytoplankton growth. On the other hand, nutrients added to the water column during the winter when light is limiting for phytoplankton growth will not necessarily result in an increase in phytoplankton growth. This means that there is the possibility for nutrients to be transported through the Kattegat during the winter months without having any effect on phytoplankton growth.

Thus, the use of an annual budget for total nitrogen delivery to/removal from a given water body as a tool for identifying the causes of eutrophication effects in coastal marine waters is limited both by the fact that most phytoplankton preferentially utilize inorganic nitrogen forms and the lack of seasonal resolution. Such budgets are, however, useful for describing the overall nutritional status of an area and for identifying long-term changes in the relative importance of various nutrient sources to a given area.

Another approach that may be applied when looking for a potential cause for an observed change, is to consider any *changes* that may have occurred concurrently in the ecosystem. In section 12.3, it has been argued that the most likely cause of the observed increase in frequency and intensity of hypoxia events in the Kattegat is a change in nutrient availability for phytoplankton. Thus, identification of any changes in the input of nutrients to this region is of importance in explaining the observed eutrophication effects.

From Table 12.1, it is clear that the major source of nitrogen and phosphorus to the Kattegat is transport from neighboring water bodies. Thus, it is important to determine whether or not there has been a major change in nutrient input via this source during recent decades which might explain the appearance of the eutrophication effects observed in the Kattegat. There is no comprehensive data series for total nitrogen and phosphorus concentrations in the waters entering and leaving the Kattegat. However, Smith and Richardson (unpublished data) have examined the available data for nitrate and phosphate concentrations in surface and bottom waters at the borders of the Kattegat during the period 1970–1992 (Figure 12.3). Note that these workers examined the borders of the Kattegat proper. This means that the southern border is placed between the Kattegat and the Belt Seas and not at the entrance to the Baltic. Thus, the surface water entering at this border represents the combined outflow from the Baltic and Belt Seas.

The data are less complete for the beginning of the period than in recent years and rigorous statistical identification of trends in the data is difficult. Nevertheless, these workers conclude that there is no evidence for a consistent change in phosphate or nitrate concentrations in bottom waters in the northern Kattegat (i.e. water entering from the Skagerrak). Likewise, there appears to be no consistent trend in phosphate concentrations entering the southern Kattegat (surface water entering the Kattegat from the Belt Seas) during the study period. There may, on the other hand, have been an increase in nitrate concentrations in surface waters entering the southern Kattegat. However, this trend is difficult to identify owing to the paucity of the data in the early years of the study period.

Assuming simple two-layer "estuarine circulation" and conservation of volume and mass of salt, Smith and Richardson (unpublished data) also examined water exchange (both horizontal and vertical) in the Kattegat on the basis of knowledge of salinity and fresh-water input. In this analysis, they were unable to demonstrate a consistent change in the amount of water entering (water exchange through) the Kattegat which may have increased the nutrient delivery from surrounding waters during the period 1970–1992. Thus, this study suggests that there has not been a major change in inorganic nutrient delivery to the Kattegat via advection from the Skagerrak (i.e. the largest "gross" inorganic nutrient contributor to the Kattegat, Table 12.1) in this period. While there may have been an increase in nitrate concentrations entering the Kattegat from the south, this study does not indicate whether these changes have occurred within the Baltic or Belt Seas.

Concentrations of nitrate in surface waters exiting the Kattegat to the Skagerrak may also have increased during the study period. Thus, there appears to have been an increase in inorganic N concentrations in water exiting the Kattegat/Baltic complex but not in the waters entering this system from the Skagerrak/North Sea. This study does not address the potential change in transport of organic nutrient sources across the borders to the Kattegat.

An increase in input of organic nitrogen to the Kattegat from the Baltic might be expected given the fact that total nutrient input to the Baltic has also increased since the 1950s [Larsson et al., 1985]. For various reasons (including the relatively long residence time for water in the Baltic and the permanent water column stratification which leads to nutrient-poor surface waters), it is argued that inorganic nutrients in the surface waters of the Baltic will be quickly utilized by the local phytoplankton population and not be transported to the Kattegat [Richardson and Ærtebjerg, 1991]. The relative proportions of organic and inorganic nitrogen in waters entering the Kattegat from the Baltic (Table 12.1) support this argument. Thus, there is some reason to suspect that the possible increase in nitrate concentrations in the water entering the Kattegat over its southern border may be the result of processes occurring in the Belt Seas.

There is a much greater percentage of inorganic relative to organic nitrogen entering the Kattegat through the Skagerrak border than from the Baltic (Table 12.1). In addition, considerably more nutrients are transported over this border than over the Baltic border (Table 12.1). Thus, input over this border may be expected to have the greatest influence on phytoplankton growth of any of the nutrient sources to the Kattegat. Again, however, Smith and Richardson (unpublished) were unable to identify a long-term change in inorganic nutrient delivery over this border (Figure 12.3).

Nevertheless, it should be noted that a number of authors have argued that the Jutland Coastal Current (JCC) may, under some conditions, transport large amounts of nutrients from the German Bight (which is under the influence of Elbe, Rhine and Weser River outflows) to the Skagerrak [e.g., Aure et al., 1990; Maestrini and

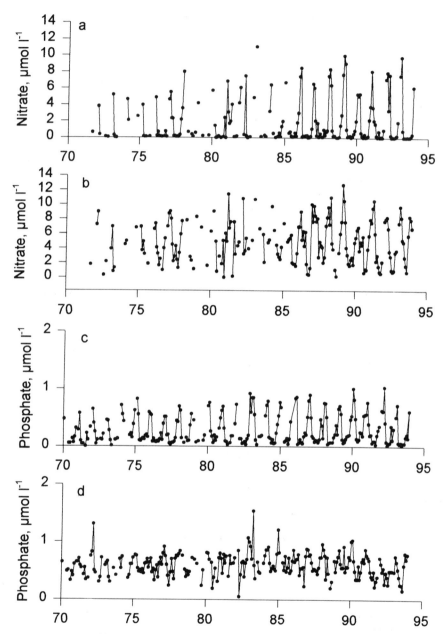

Figure 12.3A. Monthly averages in surface (0–6 m) and bottom (defined as the deepest record in all data profiles that are > 19 m in depth) nitrate and phosphate concentrations at stations sampled within the geographic coordinates of 57°20'–57°45'N and 10°20'–11°40'E during the period 1970–1993. Data from the International Exploration of the Seas (ICES) Oceanographic Database.

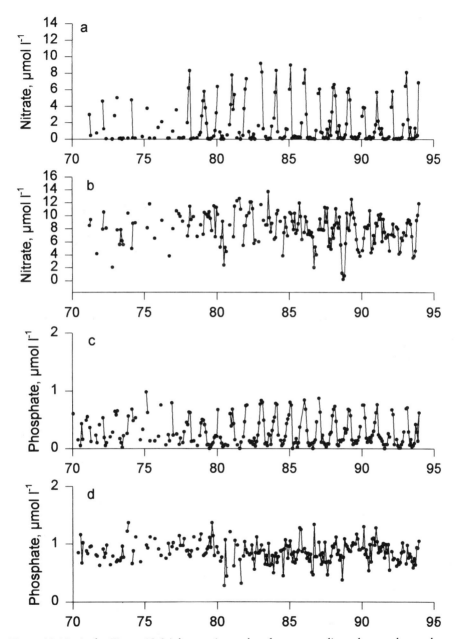

Figure 12.3 B. As for Figure 12.3 A but stations taken from a coordinate box at the southern border of the Kattegat (56°05'–56°45'N and 12°30'E).

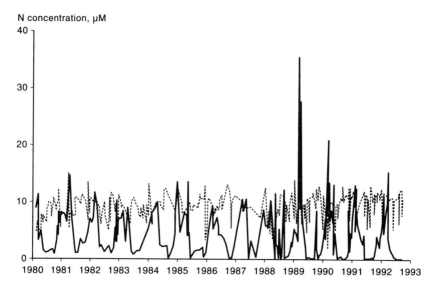

Figure 12.4. Distribution of nitrate concentrations during the period 1980–1992. -----: open North Sea water in the Norwegian trench (mean nitrate concentration in the depth 100–400 m) and ——: surface waters (mean 0.30 m) in a coastal region off the port of Hirtshals on the Danish west coast. The coastal data were selected to only include waters affected by coastal fresh-water inflow (i.e. Jutland Coastal Current) by excluding values from waters with salinity >34 psu. A seasonal pattern in nitrate concentrations is clearly evident in the coastal surface data but not in the deep Norwegian Trench Waters. From Heilmann et al. [1991].

Granéli, 1991]. It has also been suggested that these nutrients may be further transported into the Kattegat and that this nutrient source may be important in the provocation of eutrophication effects in the Kattegat. Indeed, there is at least one well-documented record in which nutrient-rich JCC water has been tracked well into the Kattegat [Ærtebjerg, 1990].

While such transport events can be dramatic, they appear to be highly episodic (Figure 12.4). The importance of such episodic events in terms of the nutrient dynamics of the Kattegat as a whole has yet to be quantified. It should be noted, however, that most of the water entering the Skagerrak and subsequently reaching the border of the Kattegat is not of JCC origin but rather originates from the open North Sea. It has been suggested that on the order of 10% [Jakobsen et al., 1994] of the water entering the Kattegat from the Skagerrak originates from German Bight Water transported by the JCC. If we assume that 10% of the water inflow to the Kattegat also contains 10% of the nutrient input then, in order to effectuate a 10% increase in the total annual input of nutrients to the Kattegat over the Skagerrak border, the concentrations in the JCC water would have had to have approximately doubled. Nitrate concentrations in the region off the north-west Danish coast which is affected by the JCC when it is present are most often at the same level as or lower than (Figure 12.4) those observed in the waters of the Norwegian

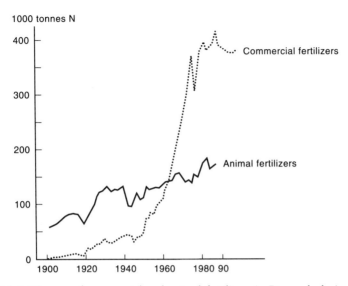

Figure 12.5. The use of commercial and animal fertilizers in Denmark during the period 1900–1989 [Danish Environmental Protection Agency, 1991].

trench (i.e. waters affected by the open North Sea and North Atlantic) which also feed the Skagerrak. Data are not available to assess whether or not changes may have occurred over time in the organic nutrient input across the Skagerrak/Kattegat border.

Thus, it seems unlikely that changes in inorganic nutrient input capable of supporting the observed increase in primary production have occurred during recent decades at the borders of the Kattegat and its surrounding seas. It is unclear for both the Skagerrak and Baltic borders whether or not changes have occurred in the transport of organic nutrient material. However, even if they have occurred, it seems unlikely that their magnitude would have been sufficient to have caused the recorded changes in primary production.

On the other hand, increases in nitrogen input via runoff and atmospheric deposition have been documented to have increased during this period [Hansen et al., 1995]. It has been suggested that changes in agricultural practice, in particular a dramatic increase in the use of commercial fertilizers (Figure 12.5) may be largely responsible for this increase [Danish Environmental Protection Agency, 1991].

12.5. Monitoring for Eutrophication Effects

Evidence for eutrophication is most often sought in changes in nutrient concentrations in the area being studied. Thus, monitoring programs are designed to identify changes in nutrient concentrations resulting from eutrophication. Often, monitoring is limited to measurement of inorganic nutrients. As pointed out in the previous section, however, it is important that consideration also be given to organic nutrient concentrations.

There is a problem in trying to identify the extent of eutrophication by examining nutrient concentrations in the water column. Basically, this problem arises because of the fact that one is trying to relate a state variable (nutrient concentration) to a change in the rate of flux (of nutrients) to the environment.

An analogy here would be trying to compare the size of bank accounts with the account holders' salaries. It is often but not always the case that the holders of the largest bank accounts are also those with the largest salaries. The rate at which money is being removed from the account is, however, also important in determining its size at any given time. The same is true with regard to nutrient concentrations in the marine environment. In fact, eutrophication effects are only observed when the incoming nutrients are incorporated into organic material. Thus, if nutrients are used quickly upon entering the marine environment, there is the possibility that a change in the rate of delivery to the system will never be reflected in the concentrations observed in the water column.

Monitoring programs often attempt to overcome this problem by focusing primarily on the winter months when phytoplankton activity is minimal and not likely to be limited by nutrients. Thus, it is argued that nutrient concentrations during winter months will be those which are most likely to exhibit changes resulting from changes in input. For a region such as the Kattegat, however, measuring nutrient concentrations during the winter and assuming that these nutrients are directly related to the phytoplankton activity and eutrophication effects recorded in the summer and autumn is problematic. This is because water exchange in the region is so variable. During some years, it occurs so quickly that the nutrients measured during the winter do not remain in the system long enough to affect production processes later in the year. During other years, the nutrients measured during the winter months provide a good indication of the nutrients available for phytoplankton growth later in the year.

Eutrophication can, then, theoretically occur without necessarily resulting in measurable changes in inorganic nutrient concentrations in the water column.

Most monitoring programs acknowledge the problems associated with using nutrient concentrations as an indicator of eutrophication and attempt to take account of nutrient "use" by including measurement of phytoplankton biomass (chlorophyll) and/or phytoplankton primary production. Chlorophyll concentration represents another state variable of the system. Thus, the problems associated with using chlorophyll determinations in assessing the condition of the ecosystem with respect to eutrophication are similar to those confronted when using nutrient concentrations. Chlorophyll measurements can be used as a proxy for biomass present at any given time. However, they tell little about the production and loss rates of biomass within the system.

In theory, primary production measurements which describe the conversion rates of CO_2 to organic material in the system ought to be a better tool than biomass measurements in identifying eutrophication effects. Unfortunately, however, most of the methods employed today for determining rates of primary production measure total production and are unable to differentiate between "new" and "regenerated" (sensu Dugdale and Goering [1967]) production (see chapters 1, 4 and 5).

The ratio of new production (that leading to a net increase in organic matter in the system) and regenerated production (based on recycling of nutrients within the system and not leading to a net increase in organic material) is not constant. Thus, two identical measured values of primary production may be associated with two very different net productions of organic material. Eutrophication effects such as hypoxia are the result of a net increase in the production of organic material within the system. Thus, one should ideally measure changes in new production over time in order to monitor changes in the ecosystem resulting from eutrophication.

In open marine systems, new and regenerated production are traditionally determined by measuring the relative assimilation by the plankton community of nitrate and ammonium [Dugdale and Goering, 1967] using ^{15}N. In such systems, the presence of nitrate represents an input from outside of the system (usually via upwelling). Thus, nitrate assimilation is assumed to give rise to *new* production while ammonium uptake represents regenerated production. As pointed out by Dugdale and Goering in their seminal paper, however, this approach does not readily lend itself to coastal marine systems because of the difficulties here in defining the boundaries of the system. In coastal ecosystems, it is not always possible to identify nitrate and ammonium as having originated respectively from outside and within the system. Thus, application of the ^{15}N method for determining the relative proportions of nitrate and ammonium uptake [Dugdale and Goering, 1967] does not necessarily identify new and regenerated production in coastal systems. Although there are techniques that at least partially compensate for this weakness in the method, none of them lend themselves easily to monitoring studies.

New production (net production of organic material) in aquatic ecosystems can often be approximated by determining sedimentation rates (most of the organic material produced in the system is, ultimately, deposited upon the bottom). Initially, one might wonder why so few monitoring programs related to the identification of eutrophication effects actually include measurements of sedimentation. Measurement of sedimentation rates in shallow coastal systems is, however, fraught with technical difficulties (i.e. separation of true sedimentation from resuspension, cf. chapter 6) and reliable methods suitable for routine use in monitoring programs are not readily available.

One approach that has been applied with some success in the study of stratified areas is the measurement of sedimentation rates *above* the pycnocline where measurements will not be influenced by resuspension. At present, however, the deployment of sediment traps is seldom included as a part of routine monitoring programs. An obvious conclusion here is that there is an urgent need for the development of good methods which can be used to determine the net production occurring in the photic zone in order to be able to more directly monitor ecosystem effects of eutrophication.

12.6. The Reduction of Eutrophication: Political Initiatives

Although concern over cultural eutrophication of marine waters is a relatively new phenomenon, it has quickly gained political interest. A group representing the Environmental Ministers from all countries bordering the North Sea has, for example, identified cultural eutrophication as one of the major environmental threats facing this sea [North Sea Task Force, 1994]. In a series of Ministerial Declarations, most countries surrounding the North and Baltic Seas have, in principle, agreed to reducing the cultural eutrophication of these water bodies with nitrogen and phosphorus by on the order of 50% by the early 2000s.

Denmark has been one of the most progressive nations with respect to the introduction of legislation designed to reduce cultural eutrophication and, in 1987, measures were adopted to reduce nitrogen and phosphorus input to surrounding waters by 50 and 80%, respectively. This plan called for the reduction to be carried out within five years (i.e. by 1993). The goals with respect to phosphorus reduction and nitrogen reduction from waste water treatment were essentially met within this time frame. However, the plan was not successful in achieving the desired reduction in runoff of nitrogen from agricultural land and new measures are currently being considered for the purpose of reducing nitrogen runoff from agriculture to marine areas.

These measures have been introduced on the basis of a general political and public awareness concerning the negative effects of excess cultural eutrophication and the belief that a reduction in nutrient inputs will benefit the environment. Our quantitative understanding of how cultural eutrophication affects marine ecosystems is still limited. In addition, the field of "recovery biology" (i.e. how and under what conditions does a system "return" to its pre-eutrophied state?) is not well developed for marine ecosystems. Thus, it is difficult at this time to predict precisely the effects of a 50% nutrient reduction on the environment.

The fact that these goals for nutrient reduction have been politically rather than scientifically defined has evoked criticism from a number of circles. It is, however, important to note that science is only one of the disciplines contributing to political decisions. Indeed, it has been argued that science alone often "fails the environment" in that the time required to obtain the "proof" required to establish cause and effect in a strict scientific sense will prevent the opportunity for intervention which may avert the occurrence of an adverse effect [Wynne and Mayer, 1993].

International political decisions to reduce terrestrial nitrogen and phosphorus inputs by 50% across the board also ignore the fact that not all countries contribute equally to eutrophication of international waters. For some countries which contribute only slightly to cultural eutrophication on an international scale, the goal of reducing nutrient input to the marine environment by 50% may appear unreasonable.

Gray [1992], for example, has argued the case for a more strategic plan for reducing cultural eutrophication to international waters by using the example of Norway. By his calculation, Norway currently contributes 0.5% of the cultural eutrophication of nitrogen to the North Sea. Proposed plans for reducing nitrogen and phosphorus inputs from Norway are expected to achieve reductions of 45–55% and 20–30%, respectively, and are estimated to cost on the order of $2 billion. Gray raises the question of whether such a reduction on an input that represents 0.5% of the total anthropogenic input of nitrogen to the North Sea will have a detectable effect on the ecosystem. The implication in Gray's argument is that the cost implications of such a small reduction are not justified in terms of the North Sea ecosystem as a whole.

It is important to note, however, that Gray's purpose in choosing this example is to illustrate problems related to international legislation. He does not address the question of whether or not the cost implications might be justified for other purposes than improvement of the North Sea ecosystem (i.e. improvement of local environmental conditions in fjords and near-coast regions). While Gray may be correct in his analysis of the limited potential benefits of eutrophication control in Norway for the North Sea ecosystem as a whole, a true cost-benefit analysis relating to Norwegian (or any other) eutrophication control programs would have to take all potential benefits of the adopted measures into consideration.

12.7. What Effect Will Remedial Actions Have?

Although the nutrient reduction goals described above were not identified alone on the basis of scientific advice, scientists are the obvious advisors when politicians try to anticipate the effects of remedial actions. As has been pointed out earlier in this chapter, the science of recovery biology (i.e. how a system responds to a reduction in eutrophication) is not yet well established. However, the few studies that have been carried out in which the responses of natural marine systems to a reduction in eutrophication have been recorded emphasize the fact that biological structure in marine systems is stochastic.

The example of Kertinge Nor, Denmark (chapter 10) demonstrates the stochasticity of biological systems very well. External nutrient loading to this system has been significantly reduced in recent years and nutrient release from the sediment has become important in determining the biological structure of the system. In 1991, sediment release of P and N amounted respectively to 3.3 and 0.4 times that coming from external sources. In 1992, P release again amounted to just over 3 times that coming from external sources but N release from the sediment was over twice that from external sources. This increased N release in 1992 was the result of the decay of a filamentous algal mat. The difference in N availability during the two years resulted in the development of dramatically different patterns of primary and secondary production during the two years.

Biological systems are inherently unstable and extreme events (weather conditions, algal blooms, etc.) can and do regularly change the direction in which the system is developing. This inherent instability in biological systems makes the quantitative prediction of the effect of specific actions on biological structure a formidable task indeed. Prediction of future events or conditions can only take place with the help of theoretical or numerical models (see chapter 11). Such models cannot, however, recreate the stochastic nature of the marine ecosystem. While many include terms representing the major processes occurring in the system, these are most often introduced as constants or as values changing in a constant manner. In reality, these processes often vary in their magnitude over short time scales and the factors controlling their magnitude are still only poorly understood (see, for example, the discussion in chapters 4 and 5 of the influence of wind events on sedimentation rates). Thus, quantitative prediction of the effects of a 50% reduction in the anthropogenic nutrient loading to the marine environment is difficult at the present time.

At a more general level, however, it is possible to predict that a reduction in nutrient will decrease primary production and, thus, sedimentation of organic material which will reduce the probability of severe hypoxia events. Hansen et al., [1995] have estimated the potential effect of the various political initiatives to reduce eutrophication in the Baltic and North Seas on the oxygen conditions in the Kattegat. These workers created a dynamic numerical model which includes hydrographic processes (both advective and vertical) as well as dominating processes in the nitrogen cycle (N incorporation into new production, bacterial regeneration of inorganic N from organic and denitrification). Input of N to the model occurs at the borders to surrounding seas, via atmospheric input and from land runoff (distributed around the area). The model is linked to the oxygen conditions of the area and is used to simulate oxygen conditions under variable N-loading conditions. Model verification was conducted by comparing simulated nitrogen and oxygen conditions during the 1980s with those actually observed. Although measurements for verification of the model using input conditions for the 1950s are limited, those that do exist suggest that the model can also simulate conditions at that time with reasonable accuracy.

The model was run assuming five different scenaria for nitrogen reduction to the system:

1. "1980s": assuming average loading and hydrographic conditions for the 1980s.
2. "1993": assuming the predicted loading conditions if the Danish legislation of 1987 had achieved its goal of a 50% reduction in land-based loading of N to the Kattegat/Belt Sea (see section 12.6).
3. "1996" where "1993" N loading is further reduced by assuming a 50% reduction in land-based N loading from North Sea and Baltic countries. This scenario also assumes a 10% reduction in atmospheric NH_4 deposition originating outside of Denmark.
4. "2000" improves "1996" conditions by assuming a further 7% reduction in the Danish contribution to the atmospheric NH_4 deposition (resulting from recent Danish legislation relating to the agricultural sector). In addition, a general 30% reduction in NO_x deposition is assumed.
5. "20??" where NO_x deposition is reduced by 64% (i.e. to the minimum level considered to be technically feasible).

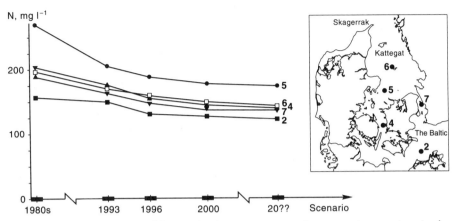

Figure 12.6. Maximum annual inorganic nitrogen concentrations at various stations in the Kattegat/Belt Seas as simulated by the model described by Hansen et al. [1995] assuming various loading conditions. See text for description of the loading conditions. Positions of stations are shown in the inset.

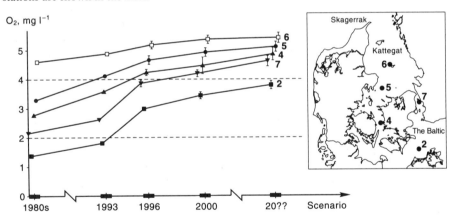

Figure 12.7. Minimum annual oxygen concentrations at various stations in the Kattegat/Belt Seas as simulated by the model described by Hansen et al. [1995] assuming various loading conditions. See text for description of loading conditions. Positions of stations are shown in the inset.

All model runs assume that the biological processes within the system have adapted to the new reduced N-loading conditions.

The maximum concentrations of inorganic N in surface waters at several stations as predicted by the model assuming the various loading conditions are shown in Figure 12.6. All of the reduction scenaria would lead to a reduction in surface N concentrations according to the model. However, the most pronounced effect occurs after a reduction in the local Danish N sources (i.e. "1993" scenario). Minimum annual oxygen concentrations in the bottom layer at these different stations were also simulated (Figure 12.7) by the model.

12.8. Time Perspective:
When Will Remedial Actions Have an Effect?

An important question in terms of eutrophication control is when remedial actions may be expected to have an effect upon the ecosystem in which these actions have been applied. Interannual variability in the factors regulating nutrient input and turnover in coastal marine system makes this question difficult to answer. Nutrient loading via runoff is, for example, a function of the amount and timing of precipitation occurring over the catchment area (Figure 12.8). In areas with restricted water exchange, the influence of local runoff (meteorological conditions) will generally be more pronounced than in more open waters where advective processes dominate nutrient delivery. Thus, year to year variability in oxygen conditions in coastal areas with limited water exchange may be greater than that observed in more open waters. This can be demonstrated by examining data on the distribution of hypoxia in the Danish Limfjord. Here, during the period from 1988 to 1994 where remedial actions have been in effect to reduce nutrient input to coastal waters, the area of the fjord affected by hypoxia has varied from less than 5 to over 30% with the extent of hypoxia being greatest in 1988 and 1994 (Figure 12.9). Thus, positive effects that may be expected to result from remedial actions directed toward a reduction of nutrient runoff may be masked by variability in the local meteorological conditions – especially those affecting winter rainfall.

Interannual variability in hydrographic patterns can also affect the extent to which eutrophication effects (i.e. hypoxia) occurs in a given year. Unusually high water exchange rates may remove nutrients from a given area before phytoplankton production and organic sedimentation take place. Therefore, the high nutrient concentrations resulting from large amounts of precipitation during winter months (cf. Figure 12.8) need not necessarily result in severe hypoxia conditions. Even when high levels of primary production and organic sedimentation take place, strong wind activity during the late summer and autumn months can reduce the probability of a severe hypoxia event by mixing oxygen into bottom waters.

Similarly, unusually calm conditions during the summer and autumn months can contribute to the development of extreme hypoxia conditions even when nutrient input/primary production/organic sedimentation have not been exceptionally high. Given the number of factors affecting the development of eutrophication effects and the inherent variability associated with all of these factors, precise prediction of when remedial actions can be expected to elicit a positive response in the ecosystem is difficult.

Another factor which complicates the prediction of when the effects of remedial actions to reduce eutrophication effects might be expected to appear is the potential for internal pools of nutrients to be stored within the system in sediments. Via this mechanism, nutrients can, in theory, be released from sediment pools and continue to stimulate high levels of primary production and organic fallout long after other anthropogenic sources of nutrients to the system have been reduced. Time series de-

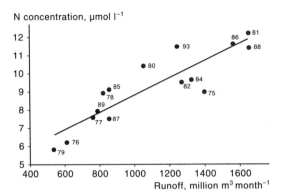

Figure 12.8. Winter concentration (mean January–February) of inorganic N in the surface layer (0–10 m) in the southern Kattegat and Great Belt as a function of winter runoff (mean December–February) from Denmark in the period 1975–89. From Richardson and Ærtebjerg [1991].

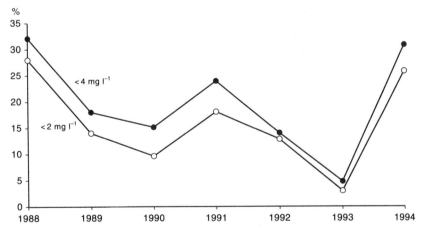

Figure 12.9. Percentage area in the Danish Limfjord annually affected by hypoxia in the period 1988–1994. Data from The County of Viborg, Denmark, Environment and Technical Department.

scribing burial of N and P in sediments are lacking for most coastal marine areas. Thus, the degree to which cultural eutrophication may have altered nutrient burial rates is unclear.

However, considerable nutrient reserves may be buried in the sediment. It is estimated (Christian Christiansen, pers. comm.) that 21,000 tonnes of nitrogen and 6500 tonnes of phosphorus are buried annually in the Kattegat proper (i.e. excluding the Belt Seas). Bo Barker Jørgensen (pers. comm.) has independently estimated nitrogen burial in the Kattegat and Belt Seas to be 25,000 t yr^{-1}. Much of this burial is recorded in the northern regions of the Kattegat [Jørgensen et al., 1990] and may be associated with sedimentation out of North Sea/Skagerrak waters after entering the Kattegat. Jørgensen et al. [1990] estimated that nitrogen burial at a station in the open region of the Belt Sea (i.e. with strong current conditions) represents on the order of 1% of the annual phytoplankton requirement in this area. On the order of 30–40% of the annual phytoplankton requirement for nitrogen may, on the other hand, be buried each year in the northern Kattegat.

Processes relating to nutrient release from sediments are still not well understood (see chapters 6 and 7). Thus, it is difficult to predict how this nutrient reservoir may counteract eutrophication control measures. It does seem clear, however, that nutrient release from sediments will also vary interannually depending on oxygen (redox) conditions, magnitude of resuspension, mechanical disturbance of the bottom (through, for example, fishing and gravel extraction activities) etc.

During the most recent years, measurements of oxygen concentrations in bottom waters at some stations in the Kattegat have been somewhat higher than those recorded during the 1980s. Proponents of the Danish eutrophication control measures initiated for these waters in 1987 argue, of course, that the apparent improvement is a direct result of these actions. However, given the inherent variability in the system, it is not possible on the basis of only a few years of data to establish with statistical certainty that there has been a persistent change in the oxygen condition of the Kattegat. There is the very real possibility, even if remedial actions have had a positive effect, that years with low oxygen concentrations during summer months will still occur.

If one assumes that the variances observed around the regression line describing the decrease in oxygen concentrations observed in the southern Kattegat over time (Figure 12.1) are an expression of the natural variability occurring in the factors regulating nutrient turnover here and that this variability will continue in the future, then it is possible to calculate how long a data time series would be required in order to demonstrate with an acceptable level of confidence that a change in the ecosystem has occurred. Let us, for the sake of argument, hypothesize that the oxygen concentrations during the late summer in the southern Kattegat will now be constant (i.e. slope = 0) rather than decreasing as a result of the remedial actions taken to combat eutrophication. In other words, let us test a constant decrease rate of oxygen depletion against a constant oxygen concentration (i.e. no further decrease over time).

Statistical analysis of this hypothesis (assuming that the level of variation about the new line will be exactly the same as that around the old line (Figure 12.1)) indicates that 21 years of data (Table 12.2) would be required in order to discriminate a continued negative gradient of 0.050 mg O_2 l^{-1} yr^{-1} (i.e. the slope observed in the line describing oxygen concentrations over the last 30 years) as opposed to a zero gradient (slope) with 95% confidence. The usually accepted significance for this type of error is 80% [e.g., Cohen, 1988] which would require 18 years of data. Being able to detect a much more rapid ratio of improvement (i.e. a gradient of ±0.100) with 80% confidence will require 14 years of data. Even detecting a gradient of ±0.100 with only 50% confidence will require 11 years of data.

TABLE 12.2. The number of years of data required to attain the power shown against various gradients (in oxygen) concentration over time in a test of zero gradient. For example, the table indicates that 21 years of data would be required in order to be 95% certain that we could detect that the gradient (slope) of the line describing the oxygen concentrations in the bottom water of the southern Kattegat during the months of August–October was 0.05 mg O_2 l^{-1} yr^{-1} rather than zero. Analysis conducted by E. McKenzie, Strathclyde Univ., UK.

Gradient	Power									
	0.50	0.55	0.60	0.65	0.70	0.75	0.80	0.85	0.90	0.95
0.005	63	66	69	72	75	79	82	87	92	99
0.010	39	41	43	45	47	50	52	55	58	63
0.050	14	14	15	16	16	17	18	19	20	21
0.100	9	9	9	10	10	11	11	12	12	14

Proponents and opponents of the remedial actions that have been taken against eutrophication effects in the Kattegat are forced to argue their cases at the moment on the basis of data taken during one or, at most, only a few years. From the statistical analysis presented above, however, it is clear that there are no immediate answers to the question of whether or not remedial actions directed toward a reduction of hypoxia in this coastal marine ecosystem have had the desired effect. Realistically, a data set of on the order of 20 years will probably be required before scientists will with reasonable confidence be able to ascertain whether or not Danish legislation enacted in 1987 has reduced the intensity of hypoxia events in the southern Kattegat.

The number of years required to obtain statistical evidence for a change in the ecosystem will vary from system to system depending upon the individual characteristics of the system. However, as a general rule for all coastal marine ecosystems, immediate quantifiable effects of measures designed to reduce eutrophication should not be expected. In most cases, many years of data will be required in order to identify with statistical certainly changes in the ecosystem.

12.9. Science and "Safe" Limits of Eutrophication

In the Introduction to this book (chapter 1), it was argued that it is the responsibility of the scientific community to develop the necessary expertise to answer the questions of politicians, administrators and the public at large concerning the effects of cultural eutrophication on the marine ecosystem and to advise concerning the relative value of various proposed remedial actions. It should be emphasized here, however, that scientists cannot be expected to identify and recommend courses of action to combat cultural eutrophication in the absence of clear goals for environmental conditions desired in a given area.

In fresh-water systems, Vollenweider-type empirical models have been employed to identify "acceptable" levels of eutrophication in specific water bodies (see chapters 1 and 11). This has often been interpreted as scientific advice concerning "acceptable" levels of cultural eutrophication. It is important to note, however, that while the concept of "acceptable"/"tolerable"/"safe" levels of eutrophication may be useful in an administrative sense, it has no real meaning in a scientific context. In order to evaluate whether or not levels of eutrophication can be considered as "acceptable", it is necessary to have defined quality criteria for the conditions desired in the environment for which acceptable limits are being discussed. Such quality objectives must be established politically. However, scientific input is required in order to identify realistic objectives and the methods that may be used to achieve them.

References

Ærtebjerg, G., Nitrate inflow from the North Sea in 1989 (in Danish), *Scientific Report from the Danish Environmental Protection Research Institute*, 4, 109–111, Copenhagen, 1990.

Aure, J., E. Svendsen, F. Rey, and H. R. Skjóldal, The Jutland Current. Nutrients and physical oceanographic conditions in late autumn 1989, *ICES CM90 C:35*, 1990.

Cohen, J., *Statistical Power Analysis for the Behavioural Sciences*, 567 pp., Erlbaum Associates, Hillsdale, New Jersey, 1988.

Danish Environmental Protection Agency, Environmental impacts of nutrient emissions in Denmark (in Danish), *Report from the Danish Environmental Protection Agency*, 1, 203 pp., Copenhagen, 1991.

Dugdale, R. C., and J. J. Goering, Uptake of new and regenerated forms of nitrogen in primary productivity, *Limnol. Oceanogr.*, 12, 196–206, 1967.

Gray, J. S., Eutrophication in the sea, in *Marine Eutrophication and Population Dynamics*, edited by G. Colombo, I. Ferrari, V.U. Ceccherelli, and R. Rossi, pp. 3–15, Olsen & Olsen, Fredensborg, Denmark, 1992.

Hansen, I. S., G. Ærtebjerg, and K. Richardson, A scenario analysis of effects of reduced nitrogen input on oxygen conditions in the Kattegat and the Belt Sea, *Ophelia*, 42, 75–93, 1995.

Hecky, R. E., and P. Kilham, Nutrient limitation of phytoplankton in freshwater and marine environments: a review of recent evidence on the effects of enrichment, *Limnol. Oceanogr.*, 33(4(2)), 796–822, 1988.

Heilmann, J. P., D. S. Danielssen, and O.V. Olsen, The potential of the Jutland Coastal Current as a transporter of nutrients to the Kattegat, *ICES C.M. 1991/C:34*, 1991.

Hinga, K. R., Alteration of phosphorus dynamics during experimental eutrophication of enclosed marine ecosystems, *Mar. Pollut. Bull.*, 21(6), 275–280, 1990.

Isaksson, I., L. Pihl, and J. Van-Montfrans, Eutrophication-related changes in macrovegetation and foraging of young cod (*Gadus morhua* L.): A mesocosm experiment, *J. Exp. Mar. Biol. Ecol.*, 177(2), 203–217, 1994.

Jacobsen, T. S., Energy production by current and wind in the Great Belt (in Danish), *Havforskning fra Miljøstyrelsen*, 31, 49 pp., Copenhagen, 1994.

Jakobsen, F., G. Ærtebjerg, C. T. Agger, N. K. Højerslev, N. Holt, J. Heilmann, and K. Richardson, Hydrographic and biological description of the Skagerrak front (in Danish), *Havforskning fra Miljøstyrelsen*, 49, 106 pp., 1994.

Jørgensen, B. B., M. Bang, and T. H. Blackburn, Anaerobic mineralization in marine sediments from the Baltic Sea-North Sea transition, *Mar. Ecol. Prog. Ser.*, 59, 39–54, 1990.

Larsson, U., R. Elmgren, and F. Wulff, Eutrophication and the Baltic Sea: Causes and consequences, *Ambio*, 14(1), 9–14, 1985.

Maestrini, S. Y., and E. Granéli, Environmental conditions and ecophysiological mechanisms which led to the 1988 *Chrysochromulina polylepis* bloom: an hypothesis, *Oceanol. Acta* 14(4), 397–413, 1991.

North Sea Task Force, *North Sea Quality Status Report 1993*, 132 pp. Oslo and Paris Commissions, International Council for the Exploration of the Sea, Olsen & Olsen, Fredensborg, Denmark, 1994.

Oviatt, C., P. Lane, F. French, and P. Donaghay, Phytoplankton species and abundance in response to eutrophication in coastal marine mesocosms, *J. Plankton Res.*, 11(6), 1223–1244, 1989.

Richardson, K., and G. Ærtebjerg, Nitrogen, phosphorus, and organic material in the terrestrial and marine environment, in *Report from a Consensus Conference* (in Danish), Danish Ministry of Education and Research, 1991.

Richardson, K., and J. P. Heilmann, Primary production in the Kattegat: past and present, *Ophelia*, 41, 317–328, 1995.

Smetacek, V., U. Bathmann, E.-M. Nöthig, and R. Scharek, Coastal eutrophication: causes and consequences, in *Ocean Margin Processes in Global Change*, edited by R. F. C. Mantoura, J.-M. Martin, and R. Wollast, pp. 251–179, John Wiley & Sons, Chichester, 1991.

Steemann Nielsen, E., The use of radio-active carbon (C^{14}) for measuring organic production in the sea, *J. Cons. Intl Explor. Mer.*, 18, 117–140, 1952

Wynne, B., and S. Mayer, How science fails the environment, *New Scientist, 138*, 33–35, 1993.

Index

269

Author Address List

Frede Østergaard Andersen
Odense University
Institute of Biology
Campusvej 55
DK-5230 Odense M
Denmark

Willem A. H. Asman
National Environmental Research Institute
Frederiksborgvej 399,
P.O. Box 358
DK-4000 Roskilde
Denmark

Jens Borum
Freshwater Biological Laboratory
University of Copenhagen
Helsingørgade 51
DK-3400 Hillerød
Denmark

Lars Hagerman
Marine Biological Laboratoratory
University of Copenhagen
Strandpromenaden 5
DK-3000 Helsingør
Denmark

Jørn Nørrevang Jensen
National Environmental Research
Institute
Frederiksborgvej 399
P.O. Box 358
DK-4000 Roskilde
Denmark

Alf B. Josefson
National Environmental Research
Institute
Frederiksborgvej 399
P.O. Box 358
DK-4000 Roskilde
Denmark

Carsten Jürgensen
County of Fyen
Ørbækvej 100
DK-5220 Odense SØ
Denmark

Bo Barker Jørgensen
Max Planck Institute
  for Marine Microbiology,
Dept. of Biogeochemistry
Fahrenheitstr. 1
D-28359 Bremen
Germany

Lars Kamp-Nielsen
Freshwater Biological Laboratory
University of Copenhagen
Helsingørgade 51
DK-3400 Hillerød
Denmark

Thomas Kiørboe
Danish Institute for Fisheries Research
Charlottenlund Castle
DK-2920 Charlottenlund
Denmark

Søren E. Larsen
Dept. of Meteorology and Wind Energy
Risø National Laboratory
P.O. Box 49
DK-4000 Roskilde
Denmark

Jacob Steen Møller
Danish Hydraulic Institute
Agern Allé 5
DK-2970 Hørsholm
Denmark

Katherine Richardson
Danish Institute for Fisheries Research
Charlottenlund Castle
DK-2920 Charlottenlund
Denmark

Hans Ulrik Riisgaard
Institute of Biology
Odense University
Campusvej 55
DK-5230 Odense M
Denmark

André W. Visser
Danish Institute for Fisheries Research
Charlottenlund Castle
DK-2920 Charlottenlund
Denmark